PERSONAL COMMUNICATIONS SYSTEMS APPLICATIONS

Feher/Prentice Hall Digital and Wireless Communications Series

Carne, E. Bryan. *Telecommunications Primer: Signal, Building Blocks and Networks*

Feher, Kamilo. *Wireless Digital Communications: Modulation and Spread Spectrum Applications*

Garg, Vijay, Kenneth Smolik, and Joseph Wilkes. *Applications of CDMA in Wireless/Personal Communications*

Garg, Vijay and Joseph Wilkes. *Wireless and Personal Communications Systems*

Pelton, N. Joseph. *Wireless Satellite Telecommunications: The Technology, the Market & the Regulations*

Ricci, Fred. *Personal Communications Systems Applications*

Other Books by Dr. Kamilo Feher

Advanced Digital Communications: Systems and Signal Processing Techniques

Telecommunications Measurements, Analysis and Instrumentation

Digital Communications: Satellite/Earth Station Engineering

Digital Communications: Microwave Applications

Available from CRESTONE Engineering Books, c/o G. Breed, 5910 S. University Blvd., Bldg. C-18 #360, Littleton, CO 80121, Tel. 303-770-4709, Fax 303-721-1021, or from DIGCOM, Inc., Dr. Feher and Associates, 44685 Country Club Drive, El Macero, CA 95618, Tel. 916-753-1738, Fax 916-753-1788.

PERSONAL COMMUNICATIONS SYSTEMS APPLICATIONS

Dr. Fred J. Ricci

Virginia Tech

To join a Prentice Hall PTR
Internet mailing list, point to
http://www.prenhall.com/register

PRENTICE HALL PTR
Upper Saddle River, New Jersey 07458
http://www.prenhall.com

Library of Congress Cataloging-in-Publication Data

Ricci, Fred J., Dr.
    Personal communications systems applications/Fred Ricci.
       p.   cm.—(Feher/Prentice Hall digital and wireless
communications)
    Includes bibliographical references and index.
    ISBN 0-13-255878-5
    1. Personal communication service systems. I. Title.
II. Series.
TK5103.485.R53   1997
621.3845--dc20

96-42322
CIP

Production Editor: *Kerry Reardon*
Acquisitions Editor: *Karen Gettman*
Cover Designer: *Lido Graphics*
Cover Design Director: *Jerry Votta*
Marketing Manager: *Dan Rush*
Manufacturing Manager: *Alexis R. Heydt*

 ©1997 Prentice Hall PTR
Prentice-Hall, Inc.
A Simon & Schuster Company
Upper Saddle River, New Jersey 07458

The publisher offers discounts on this book when ordered in bulk quantities.
For more information contact:

Corporate Sales Department
Prentice Hall PTR
One Lake Street
Upper Saddle River, NJ 07458
Phone: 800-382-3419
FAX: 201-236-7141
E-mail: corpsales@prenhall.com

Printed in the United States of America

10 9 8 7 6 5 4 3 2

ISBN 0-13-255878-5

Prentice-Hall International (UK) Limited, *London*
Prentice-Hall of Australia Pty. Limited, *Sydney*
Prentice-Hall Canada Inc., *Toronto*
Prentice-Hall Hispanoamericana, S.A., *Mexico*
Prentice-Hall of India Private Limited, *New Delhi*
Prentice-Hall of Japan, *Tokyo*
Simon & Schuster Asia Pte. Ltd., *Singapore*
Editora Prentice-Hall do Brasil, Ltda., *Rio de Janeiro*

This book is dedicated to my family:
Mary Jo, Ferdinand, Dante, Kimberley, and Susan

Contents

Preface

The ability of people to communicate is evolving to the point where we will all have our own personal communication device. The home, office place, and factory are changing to the extent that it is not unusual to use wireless telephone, cellular phones, or intelligent pagers.

Worldwide, transparent communications to the user are beginning to proliferate. It is not unusual to have telephones in airplanes, trains, trucks, ships, etc. As personal communications systems (PCS) evolve, a person will be able to communicate from any location in the world to any other location. Someday, the Dick Tracy wristwatch—with voice, data, and video communications—will be a reality. The trend is for a person to be given a telephone number for life and receive communications wherever they are in the world.

Personal Communications Systems (PCS) and their associated devices mean different things to different people. For example, the hand-held devices used by the agricultural community for position location in the Global Positioning System (GPS) have been very effective in helping during the planting and harvesting seasons. GPS is a good example of a system utilizing a small, hand-held device that communicates with a satellite system.

United Parcel Service (UPS) uses a hand-held device to register the customers signature and send *data* back to a central computer to record the delivery of a package. This is an example of a personal communications device that makes delivery operations more efficient. Wal-Mart is beginning to implement an inventory system that uses personal communications devices to help in the stocking of shelves. In the future, there will be innumerable uses for PCS. The cellular system is growing by millions of subscribers every year; this growth is also expected to occur for the PCS.

Many books have been written about cellular systems. However, very few books have been written about the emerging field of Personal Communications Systems (PCS). This book represents a collection of work on the state of the art of PCS applications. The material was developed in the process of teaching and performing research in PCS. The emphasis of the book is on new PCS applications for the home, office, and other work places. Particular emphasis is placed on state-of-the-art PCS applications such as Low Earth Orbit (LEO) satellites, wireless LANs, adaptive cellular data systems, spectral efficiency, etc. Discussion of the new PCS standards and their implications is also included.

This book was developed to provide for state-of-the-art engineering applications in the design and development of Personal Communications Systems. Many of the engineering approaches to analog and digital system design of cellular systems are well known and written about. This book emphasizes the techniques necessary to design future systems that will utilize newly approved bandwidths in the 900 MHz and 1850–1990 MHz range. Therein lies the challenge of designing "new" systems for the newly allocated bandwidths.

The work presented in this book was undertaken out of an interest in one of the fastest growing areas of communications. In the most recent FCC auctions, over 7 billion dollars has been bid by corporations to gain access to specific PCS markets. These new markets will require exciting new technological breakthroughs and developments to meet the demands of the marketplace. The motivation for applications presented in this book resulted from the challenge for "new" technical approaches to meet the demands of additional spectrum allocation. This book is organized to enable the reader to appreciate and understand the technological approaches necessary to develop the Personal Communications Systems (PCS) of the future.

The work presented is a result of research done on Personal Communications Systems (PCS) by the Virginia Tech laboratory in Northern Virginia. Many graduate students have contributed to this work; my job was more as editor and facilitator. Included below are brief descriptions of the chapters and the graduate students who should be acknowledged for preparing the work of the various chapters.

CHAPTER

1 *Introduction to Personal Communications Systems (PCS)*
Jamie A. Guerrero
Presents an overview of Personal Communications Systems (PCS) and indicates the new frequency allocations.

2 *Overview of Cellular Systems*
Robert Clark, Peter Young
Provides an understanding of the first- and second-generation cellular standards involving analog and digital systems.

3 *Unlicensed Personal Communications Services (UPCS) Devices*
Arthur Light
Discusses a key part of the PCS frequency allocations concerned with unlicensed Personal Communications Systems (UPCS).

4   Spectral Efficiency
Lisa Mount
Discusses the important topic of bandwidth efficiency by considering comparison of
Time Division Multiple Access (TDMA) and Code Division Multiple Access (CDMA).

5   Microcell Design in a PCS Environment
N. A. Bromenshenkal
Discusses how to design the PCS systems of the future by means of PCS microcell
design.

6   The Hand-Off Problem in Cellular Radio Systems
Chris Christodoulou
Discusses the important problems of hand-offs, which will be significant in designing
new PCS systems using microcells.

7   Wireless LANs
Gary Thomas
Discusses a specific application for wireless LANs.

8   Adaptive Mobile Network
Rob Kahr
Discusses a new adaptive mobile network that will handle laptop computer and
Internet access.

9   Speech Encoding of Voice in Digital PCS Systems
Barry Hemmerdinger
Discusses the problem of speech encoding for digital cellular systems. The motivation
is to obtain the best possible spectral efficiency.

10   Antennas and Power for Mobile PCS
Gregory Gibbs
Discusses antennas for mobile cellular systems. In the future, smaller, higher-gain
antennas are needed.

11   PCS Channel Propagation in Maritime Environments
Carl Darron
Discusses how to implement PCS systems in maritime propagation environments.

12   Satellite-Based Mobile Communications Systems
Arvin Khanna, Renee Lyons.
Discusses the important aspects of satellite-based mobile communications systems,
with an emphasis on applications of Low Earth Orbit (LEO) satellite systems.

13   LEO Signal Processing Design for Telestar I
Andrea Zimbler, Michael Hustead, Krishan Bhatnagar
Discussion of a LEO Satellite to send e-mail.

It is my hope that the reader will find this book to be useful in these Engineering endeavors. No book can cover all the material in any one area. The goal of this book is to provide the reader with state-of-the-art PCS applications as a guide to the design and development of the future.

PERSONAL COMMUNICATIONS SYSTEMS APPLICATIONS

Chapter 1

Introduction to Personal Communications Systems (PCS)

It has been often said that we are on the verge of a revolution in mobile communications. This revolution will ultimately liberate us as communication users from being tied down to a particular fixed location in the telephone network and will provide us with an advanced voice and data communication capability in a highly portable package and at a reasonable price. Today, the revolution has not only started but has gained momentum. Mobile systems are fast becoming a part of our daily lives.

At the turn of the 1990s, the local telephone system of the United States was mostly wire-based. Although wireless communications had been in commercial use for several years, wireless technologies have received increasingly greater attention by consumers and industry. Wireless is in the process of unprecedented expansion that has been made possible by technological innovations into mainstream telecommunications networks that were formerly almost exclusively wire-based. Wireless communications have the potential to rapidly become the dominant form of telecommunications.

This book provides applications for the emerging personal communications services that will be established around the world. Specifically, state-of-the-art Personal Communications Systems (PCS) applications are presented as a result of research recently performed. The applications have been carefully chosen to enable the practicing engineer or student to gain a working knowledge of PCS.

1.0 WHAT IS A PERSONAL COMMUNICATIONS SYSTEM?

A Personal Communications System (PCS), defined in the broad sense, is "a set of capa-bilities that allows some combination of personal mobility and service management." We generally think of PCS as a mobile telephone service characterized by low-cost pocket telephones with service that is associated with a person instead of a place or a vehicle. Such a service could provide different services, such as voice, data and fax, to its users.

It is reasonable to expect that the majority of personal communications terminal devices will be lightweight, hand-held portable units, as illustrated in Figure 1.1. The basic requirements for a personal communications network are 1) the users must be allowed to make calls wherever they are, 2) the service must be reliable and of good quality, and 3) it must offer a range of services that the users need, such as voice, data, fax, paging and video.

It is expected that 40 percent of all telephone subscribers will be wireless by the year 2000 or shortly thereafter. Personal communications are expected to grow into a truly worldwide communications network. The economies of scale will eventually bring down the cost of the PCS available through using much wider bandwidths.

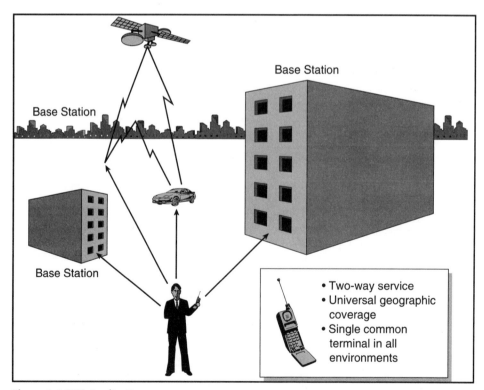

Figure 1.1 PCS Applications

1.1 WIRELESS COMMUNICATIONS: OVERVIEW OF TERMS, TECHNOLOGIES AND SERVICES

This section provides an overview of new wireless communications and a common ground for the discussion in later chapters. Wireless is an umbrella term covering many kinds of communications. Table 1.1 classifies wireless communications according to two attributes—mobility and distance of communication.

Each classification plays a different role in a telecommunications network. For example, long-haul fixed wireless communication systems, such as point-to-point microwave and satellite links, can be used as transmission media for long-distance telephone. Cellular telephony, on the other hand, serves short-haul mobile communications requirements. Other wireless communication systems can perform various functions in a telecommunications network as well. Some examples of short-haul mobile wireless communications are described in Table 1.2.

In recent years, short-haul mobile wireless services have received much attention. The interest is driven in part by the nature of wireless communication: Users are freed from mechanical connections to communication devices via wires. Technological innovations makes this type of service cheaper and better.

1.1.1 CLARIFICATION OF COMMONLY USED TERMS

Terms such as wireless, portable, mobile and cellular are often used in public discussions of wireless communications as if they were interchangeable. In reality, each has a different connotation. The following discussion clarifies terms used in this book.

- **Wireless:** Wireless describes the format of communication. Generally, wireless communications include all means of transmitting messages without a wire. This includes satellite, point-to-point microwave, cellular radio, etc.

TABLE 1.1 Classification of Wireless Systems

Mobility	Distance Short-haul	Long-haul
Fixed	Wireless local drop	Satellite relays
	Wireless PBX	Point-to-point microwave
Mobile	Cellular telephone	Land
	Cordless telephone	* Iridium
	Radio connections	Maritime
		* Inmarsat
		Aeronautic
		* Airfone

Table 1.2   Short Haul Mobile Services

Service Type	Technology	Function
Cellular telephone	Cellular radio * analog * digital	Primarily vehicular, increasingly portable
Cordless telephone	Radio connection * CT-I * CT-2	Portable
Land mobile radio specialized and private	Single base station	Vehicular and portable
Personal Communication Services (PCS)	Microcellular architecture, Digital radio	Personal

- **Cordless:** In the telephone industry, cordless usually refers to devices used in homes or offices that require no cord to connect the handset to a base unit, which itself is connected to the landline-based local exchange network.

- **Portable and vehicular:** Portable and vehicular describe functions of telecommunications. Portable telecommunication devices allow moving of handsets from one place to another while making phone calls. Vehicular devices allow one to use a telephone in a vehicle, such as a car, even when the vehicle is moving.

- **Cellular:** "Cellular," confusingly has two meanings. It is a technology term, designating the kind of wireless radio communication that requires a cellular structure of base stations. The transmission medium is radio waves, and the underlying technology can be analog or digital. In the U.S., all cellular radio systems in commercial operation are analog systems. The current trend is to move from analog to digital systems. Cellular also denotes the type of service provided by cellular system carriers.

- **Mobile:** "Mobile" is a function term. Mobile communication enables connections during motion. Although mobility is usually easier to achieve with wireless than wire-based devices, it is not tied to wireless. For example, telephones with long cords provide limited mobility. Some wireless communications are essentially fixed rather than mobile, such as between microwave towers. Mobility is not tied to cellular.

- **Personal:** "Personal," also a function term, can describe a collection of functions or features of communication services, such as small, lightweight terminals, with the ability to communicate from a variety of places and reachable at all times and locations.

1.2 NEW WIRELESS TECHNOLOGY

Different types of wireless communication technologies have been developed and implemented for decades. Satellite communication is part of everyday life; cellular telephony has been a fast-growing sector of the telecommunications industry since its establishment in 1983. Wireless has so far occupied a separate and often complementary segment of the telecommunications market. The mainstream local telecommunications have always been wire-based. Developments of wireless communication technologies such as digital encoding schemes and microcellular network architecture have the potential to rapidly propel major changes in wireless communications.

1.2.1 Digital Modulation Techniques

Historically, the only choice for telephone, whether wireless or wire-based, has been analog format. In the case of wireless communication, each radio carrier is allowed to operate in a certain frequency range, and multiple access methods are used to accommodate several telephone channels in one frequency range. Analog systems use the frequency-division multiple access (FDMA) method which divides a large bandwidth into small channels. For example, existing analog cellular networks support more than 300 channels in the 25 MHz assigned to each carrier. The telephone conversation is frequency modulated (FM) and carried into the assigned channel. In FDMA systems, each channel supports one telephone conversation at a time. The major thrust for digital modulation technology is more efficient use of the spectrum. That is, an operator can use a range of frequency to carry more telephone calls—if digital technology is employed. Two major modulation techniques have been proposed: Time Division Multiple Access and Code Division Multiple Access.

1.2.1.1 Time Division Multiple Access (TDMA)

In TDMA systems, the whole frequency range is divided into several channels, as is in analog FDMA systems, but the conversations are converted into digital format. Several conversations, say N, are assigned to one frequency channel simultaneously. The N handsets take turns for a certain length of time, transmitting and receiving the digital signals. When one handset is transmitting or receiving, each of the other N-l handsets uses the signals that it has received during the time slot assigned to it to recreate (by a digital-to-analog conversion) the original conversations. As long as the time slots are frequent enough, human ears cannot detect that a complicated sampling and synthesis process has occurred. It is estimated that current TDMA technology promises a threefold to sixfold increase of capacity over the analog systems.

1.2.1.2 Code Division Multiple Access (CDMA)

In CDMA systems, very wide (>1.25 MHZ) channels are used. A telephone conversation is digitized and modulated by a fixed-length code (called the "spread function") and then transmitted. The transmitted signal is spread out over the whole frequency range. Many coded signals, each modulated with a different spread function, can be transmitted in the same frequency range simultaneously. The handset or base station that picks up the modulated signals uses a particular assigned spread function to decode them by means of auto-correlation. Because the spread functions are all different, one handset can decode only one coded signal and recover its original "audible" format. This means that although many conversations are transmitted in the same frequency range, one handset can still pick up the coded conversation signal designed for it. This is similar to the way a person in a crowded party room hears many different voice but listens to only one of them. It is said that CDMA technology could increase the capacity by 20-fold over the analog systems.

Another unique feature of CDMA technology is the so-called "graceful degradation." When the number of simultaneous users reaches the capacity of a traditional telephone system, no callers can be added. In contrast, CDMA allows users to be added. It does result in an increase in the noise level; however, there is no strict cutoff limit of how many callers can make phone calls simultaneously.

1.2.2 Microcellular Architecture

In a cellular system, the service area is divided into cells, each with its own transmitter, receiver and antenna, which together make up a base station called a "cellsite.". The cellular structure allows a particular cell to use frequency channels currently used in cells not adjacent to itself (frequency reuse). In most cellular systems, the average cell size is between 3 to 10 miles in diameter. In a system that would use the proposed microcellular architecture, the size of a cell can be as small as 20 feet in diameter. When the cells are smaller, the system requires lower transmission power because of the shorter distance between handsets and base stations. Lower power devices translate into smaller and cheaper handsets as well as base stations.

Smaller cells imply more cells to cover a given service area. Thus, frequency reuse is greater, allowing for more channels to be available at any one time in a given cell.

There are disadvantages too in microcellular architecture. The greater number of cells means greater number of hand-offs for a moving user. The possibility of many hand-offs makes systems with microcellular architecture less suitable for vehicular applications. In addition, the investment necessary to set up a network with microcellular architecture is considerably higher than for conventional cellular systems.

1.3 EXISTING WIRELESS SERVICES

In the U.S., PCS are rapidly emerging, but the U.S. will not be the first to offer the service commercially. At this time, many U.S. companies have developed the technologies required for PCS and are now in the testing phase. In Britain, some members of the PCS family have already been offered for commercial use. The Telepoint system and Personal Communications Network, which is the British version of PCS, are two of the services with much focus, although the Telepoint system is not presently operational in this form. A number of other PCS systems are also evolving in Canada, Europe and Asia.

1.3.1 Telepoint Service

The Telepoint system, which was developed and initially introduced in 1989 in Britain, uses advanced cordless second generation technology (CT-2) to provide access to public switched network. Telepoints are multichannel, multiline base stations that are typically located in places where people congregate, such as pedestrians zones, airports, and railways or metro stations.

In its initial realization, Telepoints did not support handoffs and allowed a user only to initiate but not to receive calls. Unlike cellular telephony and paging, Telepoints do not provide continuous coverage. That means that the user has to find a station and remain within the base coverage of the Telepoint (100–300 ft) for the duration of the call.

The Telepoint service proved to be a monumental failure. As of August 1991, fewer than 10,000 users subscribed to the service (as of 1995, there are not any users.)

1.3.2 Sprint Spectrum Service

For Christmas 1995, Sprint Spectrum offered the first PCS 1900 service in the United States. The PCS service offered employs TDMA in the 1850–1990 MHz range in the Washington, D.C. area.

The services include:

- 100 percent digital network
- Containment in one personal phone
- Answering machine and pager service in phone
- Call privacy and security by using SMART (SIM) card
- Text messaging capabilities
- Detached MENU system
- Emergency access

The Sprint Spectrum service is the first to offer all digital service in the United States, although it is restricted to the Washington, D.C. area. The service appears to be fine although the voice quality is not as good as the all analog system. However, the pioneering service does offer a glimpse of things to come. The small handset can easily be placed in a pocket, purse or briefcase and utilized in travels for business or pleasure.

1.3.3 Personal Communication Network (PCN)

PCN is a form of PCS that offers metropolitan-area portable radio telephony that could compete with cellular and public-switched telephone networks. The PCN system is the European version of PCS and comes in the form of cellular and cordless. The principal idea behind PCN is to issue a unique user identification number for each individual. By means of this number, a person can be reached at any time and at any place by a caller who, in turn, does not need to know the location of the called individual. At the present time, no universally accepted definition or design concept of a PCN exists. Nonetheless, there is a broad consensus over the various general requirements that PCN will have to meet in the future. Accordingly, PCNs are primarily designed for pedestrians, using portable phones for low-cost and moderate-rate data services for communications in various indoor and outdoor settings. Requirements regarding the services characteristics of PCN, such as blocking and voice quality, are expected to be comparable to wire services. Since PCN must always know the location of a subscriber, there is a requirement for use of network intelligence (database) to transform a logical user identification number into physical location number and route calls accordingly.

1.4 PERSONAL COMMUNICATION SYSTEMS (PCS)

The so-called personal communications system is the focus of this book. The personal communication services will be made possible by developments in wireless technologies that are currently in limited experimental use. The Federal Communications Commission (FCC) has partially allocated frequency spectrum for personal communications systems. They have the potential of effective competition in businesses, including existing wireless services and traditional land-line-based telephone networks. For this reason, the regulation of this new service presents challenges to the policy makers because how this service is introduced and regulated will shape the structure of the telecommunications industry in the future and change how "public interests" are served. This section describes the personal communication services, the regulatory aspects and industry involvement.

1.4.1 What Is Personal Communication Services?

The term "PCS" is used loosely. PCS is a generic term describing mobile and/or portable, outdoor/indoor radio communications services that could provide services to individuals and businesses and be integrated with a variety of competing networks. The Telecommunications Industry Association has suggested the following definition:

> A mobile radio voice and data service for the provision of unit-to-unit communications, which can have the capability of public switched telephone network access, and which is based on *microcellular* or other technologies that enhance *spectrum capacity* to the point where it will offer the potential of essentially ubiquitous and unlimited, untethered communications.

This section does not attempt a general definition of PCS. For the purpose of discussion, however, PCS is used to mean, operationally, the type of wireless communication that implements new digital microcellular technology and provides services with features classified as "personal," as defined in section 1.0.

1.4.2 PCS Architecture

The architecture of a PCS system essentially replicates that of a cellular network with some important differences that complicate the establishment of PCS. Figure 1.2 illustrates the structure of the local portion of a PCS network. The network comprises five main components: terminals installed in cars or carried by pedestrians; cellular base stations that relay signals; wireless switching offices for switching and routing wireless telephone calls; database of customers and other network information; connections to public-switched telephone network central office. This network structure looks like a cellular system with one major difference: The PCS has *microcellular architecture* that results in smaller size of cells and larger number of base stations in comparison with those of cellular systems.

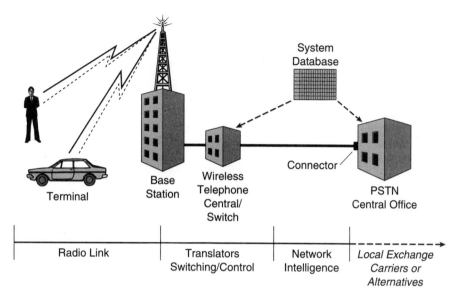

Figure 1.2 Structure of PCS Network

Technical implications include numerous hand-offs between cells for moving callers and the resulting higher processing power needed.

1.4.3 PCS Setup Cost

The microcellular architecture implies large investment to set up a PCS service. Although each PCS base station might cost less than a cellular base station because of the lower power requirement for transmission, the cost of a whole system is larger because of the greater number of base stations required.

Although the construction costs of PCS systems in the U.S. are still for the most part unknown, experiences in other markets may be used as an indicator. In the U.K., it was estimated that £4 billion (approximately $6.1 billion) was needed for setting up the digital PCN network (a British version of the PCS). If the investment is similar in scale to set up a PCS system in the U.S., few enterprises would have the financial strength to go it alone. In view of the unknown but probably high setup investment, some new patterns of competition structure may emerge to address the problem of cost. One possibility is sharing of facilities by carriers. In the U.K., Mercury Personal Communications and Unitel, two of the three PCN licensees, agreed to merge a large part of their infrastructure to reduce setup costs. It is likely that these types of alliances among competitors may emerge in the U.S. as well.

Yet another possibility is that rather than one company constructing and owning a PCS network alone, a group of companies can form a consortium in which each owns and operates a portion of the network and provides part of the service. By dividing the network, allied companies can select the portion of the network that best matches their own strengths. Each company shares the cost of setting up a whole PCS system.

Some of the stronger U.S. players in the PCS arena are listed below:

Advance MobileComm	Cox Enterprises, Inc.	Pacific Bell
American Personal Comm.	McCaw Cellular	PacTel Corp.
Ameritech	Motorola	Pagemart, Inc.
AT&T Corp.	Nationwide Wireless	Paging Networks
Bell Atlantic Comm.	Nextel Comm.	Qualcomm Inc.
Bell South Wireless	Omnipoint Comm.	Southwestern Bell

1.4.4 Regulation and Spectrum Allocation

In the U.S., the Federal Communications Commission (FCC) is the regulatory agency that administers allocation of radio spectrum. PCS, like all other radio systems, requires portions of the spectrum to operate. The problem that arises is that all of the suitable frequencies for PCS are already occupied by other services. The incumbents have already invested in the equipment operating in those frequencies, and because they would need to invest again to relocate, a high cost is associated with the reallocation of spectrum for PCS. At one time, the Cellular Telecommunications Industry Association claimed that the difference between PCS and existing services, namely cellular telephones, is too small to justify an exclusive frequency allocation.

In 1990, the FCC issued a *Notice of Inquiry in the Matter of the Amendment of the Communication's Rules to Established New Personal Communication Services* to solicit comments on the development of new wireless technologies and the establishment of new services based on them. The comments from industry indicated that the preferred frequency bands are as described below and in Table 1.3

- 800–900 MHz: This band is densely occupied by different types of land mobile services, such as private radio, specialized mobile radio and cellular telephone.
- 900–1000 MHz: Most of this band has been allocated, with the exception of three small bands: 901–902 MHz, 930–931 MHz and 940–941 MHz.
- 1,710–2,500 MHz: This band is also densely occupied. The suggested bands were: 1,710-1,850 MHz, 1,850-1,990 MHz, 2,100–2,200 MHz, 2,200–2,300 MHz and 2,452–2,483.5 MHz. They include government frequencies and various industries for point-to-point microwave communications. Those industries (petroleum, utility, railroad, broadcasting, and common carriers) strongly oppose reallocation of frequency in this range.

Table 1.3 Preferred PCS Frequency Bands

Narrowband PCS	Wideband PCS
901–902 MHz	1,850–1,990 MHz
930–931 MHz	1,910–1,930 MHz (Nationwide Unlicensed PCS)
940–941 MHz	

Source: Federal Register-r 6/24/94.

As a result, the FCC has allocated PCS spectrum as indicated in Table 1.3.

The FCC recently made history by awarding narrowband (900 MHz) licenses to the highest bidder in an auction completed July 29, 1994. Two of three megahertz were divided into eleven licenses that netted $617,006,674. The third megahertz will be auctioned at a later time. The frequency blocks that were auctioned for licensing are listed in Table 1.4.

The FCC has divided the broadband (1,850–1,900 MHz) into seven blocks: three blocks of 30 MHz, 3 blocks of 10 MHz and 1 block of 20 MHz for unlicensed PCS. These licenses will be awarded in auction(s) at a later time. The allocations are shown in Figure 1.3. *Note:* Table 1.4 is only a partial listing and is already obsolete. However, it does represent an example of what people are willing to pay for frequency spectrum.

In the most recent auction, 7 billion dollars was bid on wideband PCS. Figure 1.3 shows the old and new spectrum allocations from 1850 MHz to 2190 MHz. A portion of

TABLE 1.4 Frequency Blocks Auctioned for Licensing

Market No.	Frequency Band (M Hz)	Licensee	Cost (Millions)
N-1 (50-50 KHz paired)	940.00–940.05 901.00–901.05	Paging Networks of Virginia	$80
N-2 (50-50 KHz paired)	940.05–940.10 901.05–901.10	Paging Networks of Virginia	$80
N-3 (SO-SO KHz paired)	940.10–940.15 901.10–901. 15	KDM Messaging Co.	$80
N-4 (50-50 KHz paired)	940.15–940.20 901.15–901.20	KDM Messaging Co.	$80
N-5 (50-50 KHz paired)	940.20–940.25 901.20–901.25	Nationwide Wireless Network	$80
N-6 (50-12.5 KHz paired)	930.40–930.45 901.7500–901.7625	Airtouch Paging	$47.001001
N-7 (50-12.5 KHz paired)	930.45–930.50 901.7625–901.7750	Bell South Wireless	$47.505673
N-8* (50-12.5 KHz paired)	930.50–930.55 901.7750–901.7875	Nationwide Wireless Network	$47.5
N-9 (50 KHz unpaired)	940.75–940.80	Nationwide Wireless Network	Pioneer's Preference
N-0 (50 KHz unpaired)	940.80–940.85	Paging Network of Virginia	$37
N-11 (50 KHz unpaired)	940.85–940.90	Pagemart II, Inc.	$38

Source: FCC Public Notice #44177.

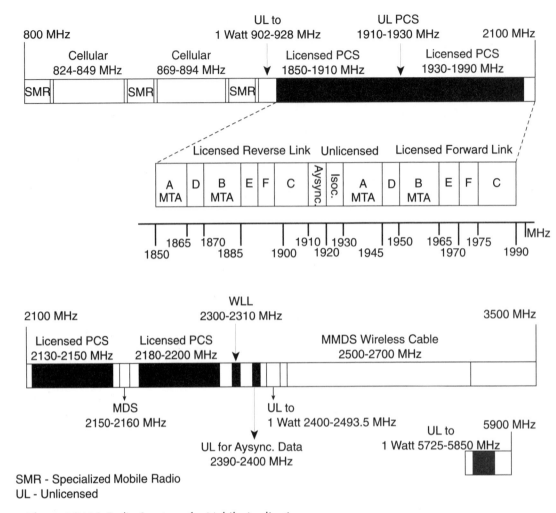

Figure 1.3 U.S. Radio Spectrum for Mobile Applications
Source: Reprinted from *Wireless Communcation*, T. Rappaport, Prentice Hall.

the spectrum will be reserved for present civilian and government use, while a portion will be put on auction for PCS use above 2.1 GHz.

FCC document number 90-134 gives the details of the PCS frequency allocations. Group 3 interested in bidding on portions of the spectrum have been developing consortiums, because the costs could be in the order of millions of dollars. It is this opening of new spectrums that is creating so much interest, since new products and services can be developed for these narrowband and wideband frequencies. In particular, new hand-held and pocket devices are currently being developed to utilize the PCS bands. Both terrestrial and satellite systems are being designed to enable the worldwide use of PCS. Low Earth

Orbit (LEO) satellite systems that cover the entire world will utilize hand-held devices that will make communication ubiquitous.

1.5 PCS TECHNOLOGY

The PCS of the future is just beginning to emerge. This book provides insight into some of the relevant design and development issues. The organization of the book provides for an understanding of the relevant technical aspects by first discussing and comparing cellular standards in Chapter 2. This is followed in Chapter 3 by a discussion of unique aspect of the PCS spectrum, namely, unlicensed PCS. The unlicensed portion of the new spectrum will allow for indoor/outdoor radio communications to Local Area Networks (LANs), private brand exchanges (PBXs), etc. This portion of the new spectrum will certainly encourage unique PCS devices for the home and business.

Before specific designs can be undertaken, one must consider the aspect of *spectral efficiency*. After all, the critical resource of spectrum must be efficiently utilized. Chapter 4 discusses spectral efficiency and gives examples for TDMA and CDMA systems. A new hybrid scheme, utilizing a combination of CDMA and TDMA, is discussed to indicate techniques of improving spectral efficiency.

Since the new PCS will utilize microcells, Chapter 5 discusses the design of microcells. A distinction in microcell design for indoor and outdoor systems is clearly made. Power requirements in future systems will be different than present cellular systems, as discussed in this chapter. As cellular sizes become smaller, the hand-off problem becomes different; Chapter 6 addresses the hand-off problem for both digital and analog systems. The details of signal strength and boundaries are discussed. In the future, communications with Local Area Networks will utilize PCS devices.

Chapter 7 presents some of the wireless LAN standards and technology. Beginning with Chapter 8, examples of new approaches to PCS are presented. Chapter 8 also discusses an adaptive mobile network that handles data communications with interfaces to the Internet. As new systems emerge, the use of improved (spectrally efficient) voice processing will be important.

Chapter 9 discusses some classical and new approaches to encoding voice in digital cellular systems. The proper design and utilization in PCS require an understanding of use of antennas on moving objects at high frequencies. Chapter 10 discusses antenna design and the effects of interference. PCS systems of the future will be placed on the sea in the air and on the ground. Chapter 11 discuses some of the propagation considerations of PCS use in maritime environments.

In the future, communications will be ubiquitous throughout the world. Individuals will be able to communicate on land, sea or air to any place at any time. In order to provide this service, the use of satellites is important. Chapters 12 and 13 discuss satellite-based communications systems with an emphasis on Low Earth Orbit (LEO) satellites. The implementation of mobile cellular communications systems on land, sea and air, utilizing spectrum efficiency, is of paramount importance in PCS. Chapter 14 discusses an application of a mobile satellite system utilizing the CDMA technique.

1.6 CONCLUSIONS

In most cases, technology developments create business opportunities and act as a driving force for economic and political changes. In the case of wireless communications, technological innovations will affect the telecommunications industry and its customers in at least two ways. First, new technologies allow companies to provide new services or existing services more effectively and at lower rates. The emergence of cellular services allows many people to use telephones in moving vehicles. Technological developments create higher capacity for cellular systems. In addition, the lower power requirement and other technical enhancements could make handsets smaller and cheaper.

Second, new wireless technologies will have a direct impact on the structure of the telecommunications industry. The existing wireless services providers, including private radio companies, paging operators, and cellular carriers, will face more competition. Wireless will also pose direct competition to the wireline networks. These competitive threats created by new wireless communications will reshape the telecommunications industry and will trigger more regulatory reforms, which will ultimately impact the industry and market structure of the future.

REFERENCES

[1] Kob, "Personal Wireless." *IEE Spectrum,* June 1993.

[2] A. Jagoda & M. de Villepin, *Mobile Communications,* John Wiley & Sons Publishers Inc., 1993.

[3] Michael Paetsch, *Mobile Communications in the U.S. and Europe: Regulation Technology & Market,* Norwood, Mass: Artech House Publishers, 1993.

[4] *Code of Federal Regulations,* The Office of Federal Register, October, 1993: 47 (part 24).

[5] *Advanced Wireless Communications,* Telecom Publishing Group, November, 1990.

[6] Milt Leonard, "PAS Are Stymied Only by Lack of Assigned Frequency Spectrum." *Electronic Design,* February 4, 1993.

Chapter 2

Overview of Cellular Systems

The personal communications systems of the future will utilize cellular techniques. Cellular technology seems like a complicated method of communication, leading one to ask the obvious question, "Why use cellular technology?" The answer lies in the limited spectrum available and continually expanding subscriber base. Cellular technology with demand–assigned channels offers the best method of packing the large number of potential users into the small bandwidth allocated. The use of cells allows frequency reuse, much as two commercial radio stations, one located in New York City and one in Los Angeles, use the same frequency. Cellular systems just place the "radio stations" closer together and use receivers that can exploit multiple radio stations. Three multiple access techniques are described and evaluated for their ability to serve subscribers on the basis of cost, efficiency, and capacity.

2.0 SYSTEM DESCRIPTION

There are three basic methods by which cellular carriers could make use of the bandwidth that they are allowed. The current system uses Frequency Division Multiple Access (FDMA) signaling method. Other proposed systems make use of Time Division Multiple

Access (TDMA) or Code Division Multiple Access (CDMA) systems. The TDMA systems can be divided into two subsets: those that first perform FDMA, then divide the resulting channels into time slots; and those that use the entire allocated bandwidth and extremely small time slots to accommodate multiple users. The CDMA systems use spread spectrum technology to allow multiple users to concurrently use the same wideband channel.

2.1 PRESENT CELLULAR ANALOG SYSTEM (AMPS)

The Advanced Mobile Phone System (AMPS) currently uses an FDMA signaling format. The portion of spectrum allocated for cellular phone use by the Federal Communications Commission (FCC) is 824 to 849 MHz for mobile transmit and 869 to 894 MHz for base station transmit, and is further divided into functions, as shown in Table 2.1. Note the fact that two carriers are sharing the bandwidth equally. [Refer to TIA AMPS standards: IS-19, IS-553, etc.]

In 1974, the FCC allocated a 40-MHz bandwidth in the 825–890 MHz frequency range for high-capacity mobile radio telephone use. This 40-MHz bandwidth is divided

Table 2.1 Frequencies Allocated to Cellular Radio (Used in AMPS) (after [1])

	A Service Provider (Nonwire Line)	B Service Provider (Wire Line)
Base cell station		
Transmit Bands (MHz)	869–880, 890–891.5	880–890, 891.5–894
Mobile station		
Transmit bands (MHz)	824–835, 845–846.5	835–845, 846.5–849
Maximum power (watts)	3	3
Cell size, radius (km)	2–20	2–20
Number of duplex channels	416*	416*
Channel bandwidth (kHz)	30	30
Modulation		
Voice	FM	FM
	12 kHz peak deviation	12 kHz peak deviation
Control signal	FSK	FSK
(voice and paging channels)	8 kHz peak deviation	8 kHz peak deviation
	10 kbps	10 kbps
	Manchester line code	Manchester line code

* 21 of the 416 channels are used exclusively for paging.

equally between transmit and receive bands. Mobile transmit channels are in the 825–845 MHz range; the mobile receive channels are in the 870–890 MHz range. This original spectrum is shown in Fig. 2.1. Channel band A is assigned to radio common carrier (RCC), and channel band B is assigned to the Wireless Common Carrier "WCC" (local telephone company).

Because of a shortage of channels, the FCC in 1989 increased the original 40-MHz frequency allocation to 50 MHz. To the original spectrum A, A' and A" were added, and spectrums B' and B" were added to B. Both wireline and nonwireline service providers have been allocated an additional 5 MHz of spectrum, divided equally between transmit and receive bands. Thus, with new spectrum allocation, mobile units transmit in the 824–849 MHz band and receive in the 869–894 MHz band, so that each full-duplex channel has a 45-MHz frequency separation. The original 40-MHz spectrum contains a total of 666 full-duplex channels to which an additional 166 full-duplex channnels have been added. The total number of channels, 832, is divided equally (416 channels each) between A (RCC) and B (WCC) service providers.

Bandwidth is assigned by the FCC for each Metropolitan Service Area (MSA) or Rural Service Area (RSA).

FDMA is one method by which the spectrum is shared. The AMPS consists of 416 full–duplex channels per service provider, for a total of 832 channels. These channels are divided among a number of cells that are designed so that adjoining cells overlap slightly, as is shown in Fig. 2.2. The number of cells among which all of the channels of the system are assigned, but none is repeated, is called a group. In the system shown in Figure 2.2, the frequencies are shared among seven cells, the number which has been shown to minimize the carrier–to–interference ratio for the AMPS system. [2]

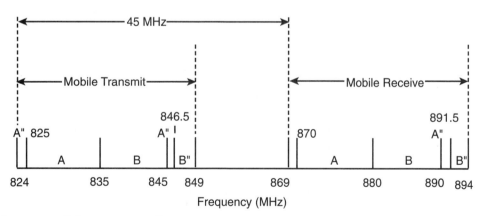

Figure 2.1 Cellular Frequency Allocation

Source: From *Cellular Radio: Analog and Digital Systems,* Asha Mehrotra, Artech, 1994. [1]

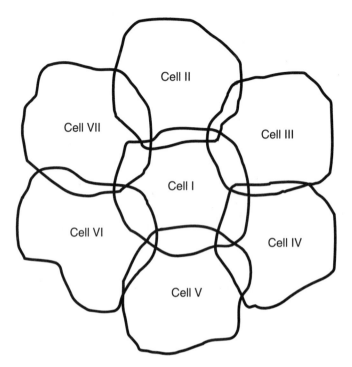

Figure 2.2 Single "Group" Cellular System

Assuming a uniform distribution of channels, each cell in Figure 2.2 would contain 59 or 60 channels. As more cells are needed to cover a wider area, the configuration would be repeated, reusing the frequencies.

The closest six copies of the seven–celled group shown in Figure 2.2 (collectively known as the first tier) are the only cells that are significant in creating cochannel interference. In a mature system, one would expect each cell to be surrounded by many other cells in a honeycomb pattern, with every seven cells completely reusing the spectrum.

2.1.2 Propagation Characteristics (AMPS)

A mature cellular system averages approximately 50 cells for an MSA. It is possible that MSAs also border other MSAs, such as the Baltimore, MD/Washington DC/Fredricksburg, VA/Richmond, VA/Martinsburg, WV contiguous service areas, at least some of which constitute different markets. As cellular systems become more commonplace, it is not difficult to envision a nationwide blanket of contiguous cells, all of which must peacefully use the same small slice of bandwidth and serve an ever-increasing customer base. This makes the antennas used, the spectral efficiency, and the capacity of the systems very important.

2.1.2.1 Typically Used Antennas

An RSA and an MSA have two different objectives. The RSA, due to the sparse distribution of users, tries to cover the maximum amount of area per cell site by use of omnidirectional antennas. The MSA, on the other hand, is faced with meeting the demands imposed by a concentrated customer base. Its antennas tend to be more specially designed to meet the needs of special cells. For those cells that focus on highway usage, the main lobe of the antenna beam is focused specifically on the highway area, since that's where the customers are. For highly urban areas with dense but distributed populations, small cells with uniform coverage are in order. These may be realized by using antennas that are focused upon either 120 degrees or even 60 degrees of a circle, as opposed to an omnidirectional antenna. The edge of a carrier's service area would be characterized by antennas that focus transmit and receive power into the carrier's territory.

2.1.2.2 Spectral Efficiency

This is a demand-assigned system, which allows the number of users to far outnumber the number of channels. Since this is an analog system, the number of bits per hertz per second is really meaningless here. A better measure of the efficiency of this system lies in the concept of the bandwidth required per user. With this system, a single user requires 30 kHz of bandwidth, 24 kHz of which is used exclusively for voice (unless blanked and used for a burst of data).

2.1.2.3 Capacity

Capacity is defined for the purposes of this book as the number of subscribers that may use the cellular system. The capacity for an FDMA system depends on a number of factors. The list of determining factors includes the following:

Loading
- The number of calls placed by the system's subscribers
- The way the calls placed are distributed in time
- The average time per call

System Layout
- The amount of frequency reuse utilized
- The size of the cells

2.1.3 Channel Attributes

The maximum allowed transmitted power of the land station is 100 watts at a height of antenna above terrain (HAAT) of 500 ft. A typical implementation, however, has a transmitter power of 20 watts and a HAAT of 100 ft. [2]

Three types of mobile radios are defined by the AMPS and approved by the FCC, and have transmit powers as summarized in Table 2.2.

Depending upon the type of antenna used at the base station, the effective radiated power can be raised by 8 or 9 dB by using more highly directional antennas in place of omnidirectional ones.

Path loss in cellular radio is highly dependent upon the type of terrain that the cell in question covers. When the signal strength in dBm and the distance from the transmitter in miles are considered, the following generalizations can be made. The lowest loss possible, the free space loss, is 20 dB per decade. The lowest smooth earth loss, approximately 35 to 40 dB per decade, can be found in an open area. A loss of roughly 40 dB per decade can be expected of a suburban area, although the loss mechanisms are generally different from those of an open area. An urban area generally has the highest loss of all due to extreme multipathing and shadowing—roughly 38 to 50 dB per decade, depending upon the layout of the city. Special loss situations such as shadowing can add even more loss to a propagation path. Depending upon the season, geographic region and density of foliage, foliage loss can increase the losses by as much as 40 dB per decade.

A general formula to compute the power received at the mobile unit [2] in a Rayleigh fading environment is:

$$Pr = Pt + Gt - 156 \text{ dB} - 40 * \log10 \ r + 20 * \log10 \ ht + 20 * \log10 \ hr + Gm$$

where:

Pr = Power received in dBm

Pt = Power transmitted in dBm

Gt = Transmit antenna gain in dBd

r = Path length in miles

ht = Height to transmit antenna in feet

hr = Height to receive antenna in feet

Gm= Antenna gain in dBd

Once the propagation model is developed, the carrier and interference (undesired) received signal levels can be determined at any mobile or base site. The ratio between the carrier (C) and the interfering (I) signal level is called the C/I ratio. A minimum C/I ratio of 18 dB is required in normal environments for acceptable performance. [2]

Table 2.2 Mobile Unit Transmit Power

Power Class	Power Level	Tolerance
I	6 dBW	8 dBW > P > 2 dBW
II	2 dBW	4 dBW > P > -2 dBW
II	-2 dBW	0 dBW > P > -6 dBW

Decibels above a watt (dBW).

2.1.4 Call Origination Sequence

Two types of call origination are possible: a call originated from a cellular subscriber to a landline subscriber, and a call originated from a cellular subscriber to another cellular subscriber. The origination sequence is the same for either, but the act of locating and connecting to the subscriber at the other end of the call is handled differently by the MTSO for each of the two cases.

When a call is originated from a cellular subscriber to another subscriber, the cellular originator first selects a telephone number and enters it into his/her mobile radio. Next, the originating user punches the "send" button, which opens a reverse setup channel to the cell site. The originating and receiving telephone numbers, as well as the information necessary to identify the mobile radio, are sent to the cell site, which then relays the information to the Mobile Telephone Switching Office (MTSO). The MTSO selects an idle traffic channel, if one is available, then notifies the cell site of the channel it has selected. The cell site forwards the information about the idle channel to the mobile radio and simultaneously turns on a 6 kHz Supervisory Audio Tone (SAT). When the mobile radio loops back the SAT to the cell site, the cell site notifies the MTSO that the channel setup has been completed. At this point, the originator has set up a channel to the MTSO. It is then the duty of the MTSO to begin the process of ringing the line of the receiving landline or cellular user. [1] Fig. 2.3 graphically shows this procedure.

Call origination for a cellular user contacting a cellular user is conducted exactly the same as call origination from a cellular user to a landline subscriber (see Fig. 2.3). The difference in the call processing is all in the MTSO's handling of the call.

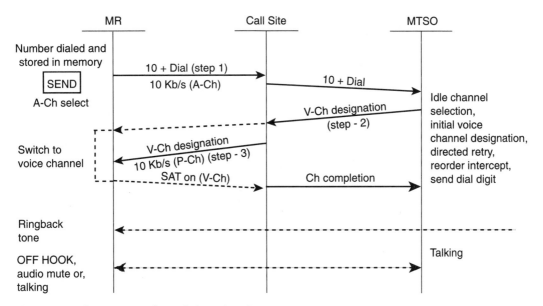

Figure 2.3 Call Origination by Cellular Subscriber
Source: Reprinted from [1].

When a landline subscriber wishes to contact a cellular user, the PSTN (public–switched telephone network) recognizes the telephone number as one belonging to a cellular phone. The MTSO notifies the cell site via a control channel that it must page the mobile radio. The cell site pages the mobile radio at 10 kbps on a paging channel. The mobile radio responds at 10 kbps on the reverse access channel, which the cell site relays to the MTSO. This lets the MTSO know that it has "found" the mobile radio and that it should set up a voice channel. Assuming a voice channel is available, the MTSO selects one, then notifies the cell site of its selection. The cell site simultaneously puts a SAT on the selected voice channel and notifies the mobile radio of the selected channel via the forward access channel. When the mobile radio has tuned to the appropriate channel and found the SAT, it retransmits the SAT back to the cell site. The cell site notifies the MTSO that a channel has been set up, then instructs the mobile radio to ring. If the ST remains on, then the call is not answered. Conversely, if the call is answered, the ST is turned off. This is the last step of the process—the users may now begin talking on the channel. [3] See Fig. 2.4.

2.1.5 Call Termination Sequence (AMPS)

Two termination sequences are of interest, that of a cellular user (CU 1) terminating the call and that of the MTSO terminating the call. We say that the MTSO terminates the call because if CU 1 does not terminate the call, it does not matter to CU 1 whether the caller who terminates the call is the landline or the other cellular subscriber: The notification and termination is conducted through the MTSO.

Termination by cellular user CU 1: The procedure for termination of a call by the cellular user CU 1 is simple. When the call has been completed, CU 1 merely presses the "end" button on the mobile radio. This causes the mobile radio to turn on the supervisory tone (ST), which, together with the SAT, indicates an "on hook" condition at the mobile radio.

The cell site then alerts the MTSO that the mobile radio is on hook. The MTSO then disconnects the channel that the communication had occurred upon and notifies the cell site to turn off the transmitter on the appropriate channel. [3] Fig. 2.5 shows the release sequence when initiated by the cellular user.

Termination by CU 1's MTSO: The termination of a call by the MTSO is just as simple as the termination by the cellular user. The on hook condition caused by the remote termination of the call is sensed by the MTSO, which sends a release code to the cell site. The cell site then forwards the release code to the mobile radio. The mobile radio then turns on the ST, which indicates the on hook condition when transmitted with the SAT. The cell site, upon sensing the on hook signal from the mobile radio, notifies the MTSO. The MTSO disconnects the channel and tells the cell site to cease transmission. [3] Fig. 2.6 shows an MTSO-initiated release.

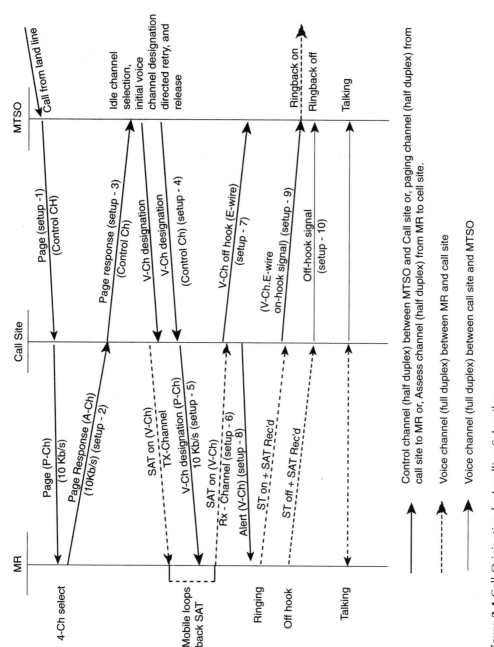

Figure 2.4 Call Origination by Landline Subscriber
Source: Reprinted from [1].

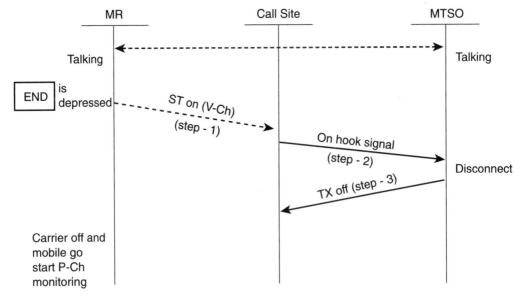

Figure 2.5 Cellular User Initiated Release
Source: Reprinted from [1].

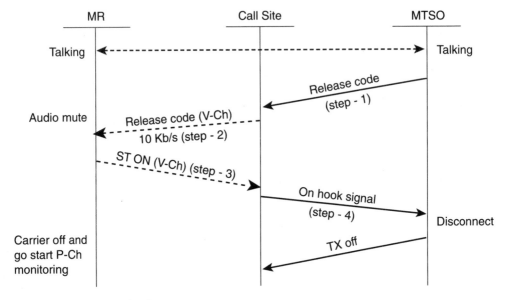

Figure 2.6 MTSO Initiated Release
Source: Reprinted from [1].

2.1.6 Description Of Hand-Off

Inside a cell are three distinct regions with different hand-off circumstances. Fig. 2.7 illustrates these regions, using cell I of Figure 2.2 as the primary cell. The central region of cell I, marked "1," is the area in which a need for a hand-off would be caused by a deep fade, but in which the signal strength of another cell is unlikely to be high enough to effect a transfer. In this area, one would probably not desire a hand-off to occur. On the outskirts of cell 1, in the regions marked "2," two cells overlap and provide approximately equal signal strengths (on average). This region is a natural one in which to effect a hand-off for a unit changing cells. However, a good hand-off routine should not continually pass a stationary user back and forth between two cells. The last regions, each marked with a "3," are regions in which the primary cell should have roughly equal signal strength with two other cells. This area will generally have three cells with the potential to hand off a cellular user rapidly among themselves.

Assuming that a cellular communications channel is in use, a hand-off becomes necessary when the signal level received from the mobile radio drops below a specified value at the Mobile Telephone Switching Office (MTSO). This prompts the MTSO to send a signal to the surrounding cell sites, "asking" them what the carrier level that they receive is. The polled cell sites respond by sending back the level of the signal that they receive from the mobile radio. The MTSO then selects an idle channel in the cell site, which responds with the highest level and a hand-off message sends to the old cell site that includes the frequency of the channel in the new cell site. It also instructs the new cell site to begin transmitting on the new channel so that the hand-off will appear seamless. The old cell site notifies the mobile radio that a hand-off is ready to take place by sending a hand-off message to it. During message transfer, the mobile radio mutes the voice channel so that the user is not disturbed by the message. The mobile radio then drops the carrier at the old cell

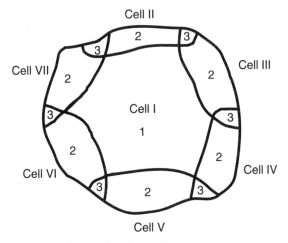

Figure 2.7 Boundaries of Other Cells Extending into a "Primary" Cell

site and tunes to the channel of the new site that was assigned to it by the MTSO for continuing transmission. When the MTSO is notified by the new cell site that the mobile radio is now using one of its channels, the MTSO tells the old cell site to stop transmitting. The MTSO can then reassign the channel which was used in the old cell if it is needed. [3] See Fig. 2.8.

There are times when hand-offs are not possible. Several scenarios are:

- When leaving an area of coverage and no cell is available
- When the only cell available is completely saturated and a channel is thus not available for assignment
- When the cell that the mobile radio enters is out of commission by virtue of a failure
- When entering an area well within cell boundaries which has a low receive signal level to and/or from the base station

The hand-off is conducted by the MTSO, the mobile radio, the old cell site and the new cell site. All of the actions and communications of these entities are transparent to the voice and low-speed data user.

2.1.7 Complexity And Cost Of System

This is the least complex system of those reviewed in this book. Other systems have been designed to improve performance in one aspect or another, but this system has several important points in its favor. The price of the mobile radio is the cost that the consumer sees first: Low cost is an important factor in enticing a customer to use a system in the first place. The AMPS mobile radio can be purchased for less than $100 in its crudest form. Some service providers go so far as to give away cellular phones when the user signs a contract to operate with that carrier for a given period of time. Most of these deals both require and provide activation at an additional cost, but the consumer likes the idea of getting something for nothing. The carrier will more than recoup its expenditure in its service charges. Depending upon the type of plan selected, cellular calling under this system can be both reasonable and affordable.

The cost of the network components (excluding mobile radios) is an important consideration for the system operator (carrier). Omnidirectional antennas are more inexpensive than directional antennas, but as antenna pattern shaping becomes necessary, the cost of the antennas goes up. If the antennas are designed to handle only a sector of a cell, the cost of the switching equipment driving the antennas becomes higher as well. However, the cost of the system must be justified by increase in capacity and, ultimately, revenues. Otherwise, the management of the carrier would never have allowed the improvements and refinements to take place. Frequency modulation is a well-understood and relatively simple transmission scheme. It requires no complex timing circuits and no complex correlation routines. Of all of the transmitting and receiving equipment, this is by far the simplest (and therefore the least expensive).

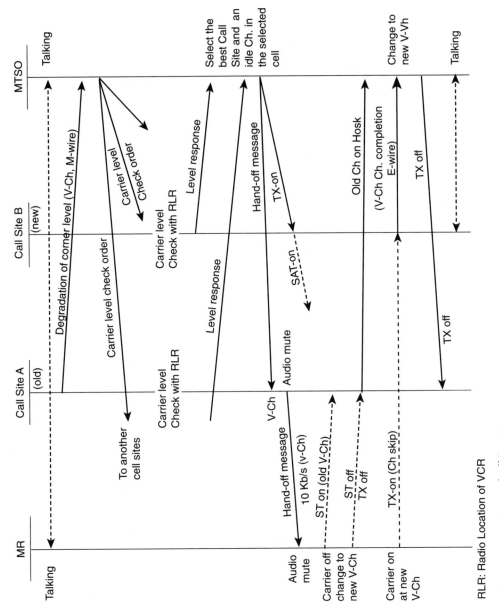

RLR: Radio Location of VCR

Figure 2.8 AMPS Hand-off Routine
Source: Reprinted from [1].

2.1.8 Dividing Channels Of The Amps System
Into Time Slots (NADC)

In references [4] and [3], reference is made to an interesting system, that uses a Time Division Multiple Access (TDMA) scheme called North American Digital Cellular (NADC). This is the first digital system that has been standardized in North America. In this system, the same 30-kHz channels of AMPS are used, but each channel is divided into three time slots. This results in a single channel being shared by a maximum of three users, for a threefold increase in the number of channels and, theoretically, a threefold increase in capacity. Of course, the number of channels that must be supported increases by three times, thus creating extra demand upon the forward and reverse setup channels. Depending upon the effect of this demand, more bandwidth might have to be allocated to this "overhead."

By and large, the propagation characteristics of the digital NADC should be the same as those for the analog FDMA. The NADC might be slightly more susceptible to the effects of short-term fading. The antennas used should be exactly the same in either system. If fading is more of a problem in NADC, directional antennas might be used to increase the gain on the link, thus providing a higher received carrier and therefore higher carrier to interference (C/I) ratio. From both the spectral and subscriber standpoint, the NADC would be more efficient than the regular AMPS system. Although the increase in subscribers and spectral efficiency might not be threefold, it should be at least twofold.

Call origination, call termination, and hand-off should occur largely in the same manner with some important exceptions:

- The number of channels increases by a factor of two or three, meaning that the switching equipment at the cell site and the MTSO must be able to handle the additional volume.
- The overhead associated with dynamically assigning these channels could exceed the capacity of the setup system designed for the AMPS. More channels might be needed, which would necessitate an algorithm for selecting setup channels.
- Hand-offs could occur between time slots within the same channel, in addition to the hand-offs between cells and between channels.

As was already implied, the complexity of the NADC system increases in terms of call handling and switching capability at the cell site and at the MTSO. The antennas are largely of the same complexity as those used in the AMPS, but the gain acquired by using directional antennas becomes more attractive. Both the transmitters and receivers become more complex, as timing circuits are needed to ensure that a transmitter does not infringe upon another time slot within the channel. Ultimately, adding capacity becomes more expensive for both the cellular carrier and for the cellular user. This added cost is offset to the carrier by the added number of subscribers that provide it revenue, but in the end the cellular user must bear the cost.

2.2 TDMA STRUCTURE—GSM

With the proliferation of computers, digital communication technologies have advanced rapidly. Ten years ago, the AMPS system was thought to be the system that would carry the U.S. well into the 21st century—a digital system was impractical and unneeded. The growth of high-speed digital signal processing has caused a major revolution in the communications industry. To consumers, the word "digital" means high quality and low noise because of the wide acceptance of the compact disc and digitally recorded music. So, today, the trend is toward digital cellular technologies.

The TDMA structure allows multiple users to share bandwidth by giving a slice of time to each user for transmitting and receiving data. In a pure TDMA system, the user's data (voice or computer data) and the overhead control data is transmitted in one slice and received in another. One of the inherent complexities of this system is synchronization. Each mobile unit must time its transmission so that it arrives at the base station at the correct time. Accurate distance and time-delay measurements are required in order to compute the correct transmission time or time advance. In addition, the mobile unit is usually moving, requiring measurements frequently enough to prevent the mobile user's data from arriving during an adjacent time slot.

There are several advantages of using TDMA:

* Burst mode transmission: The mobile unit transmits only during its time slot, resulting in lower battery power consumption. This power savings can result in either increased usage time or reduced weight.
* Increased quality of the voice channel: The voice coding is generally at a high enough data rate that bit errors do not cause significant degradation of the voice channel.
* Secondary benefits: Since TDMA is a digital coding technique, overhead bits can provide for measurements and aid in making the hand-off more seamless.

In a practical implementation, the TDMA architecture must be a mix of TDMA and FDMA. In the Group Special Mobile (GSM) standard, the spectrum is broken up into channels like FDMA, and multiple users access each channel, using TDMA. In Europe, a large conversion process is in progress to migrate their current analog system to the GSM standard. The current analog systems being used throughout Europe vary from country to country, making it harder for a single user to roam the European community with one phone. The intent of the GSM standard is to unify the European community by offering a single standard so that users throughout Europe can use one phone. Current estimations show that the number of GSM users will exceed the number of European analog users by 1996. [1]

2.2.1 Background—Basic System Description

The GSM standard was designed with the open systems interconnection (OSI) model for computer communications in mind. This allows system operators to implement different features to optimize their system for their particular environment. This flexibility includes the hand-off algorithm, cell selection, power control, and call dropping parameters, to name a few. These parameters can be modified by the system engineers to fit system needs while keeping one standard mobile phone.

There are 125 duplex channels in the GSM standard accommodating eight users on each channel. Up to eight users on a given channel are assigned different time slots for transmission of data. Using Gaussian minimum shift keying (GMSK), the overall data rate through the 200-kHz-wide channel is 270.833 kbits/sec. GMSK is a Gaussian filtered continuous phase, frequency shift keyed (FSK) signal with a minimum modulation index (0.5) that produces orthogonal signaling. This is a bandwidth conservation technique that has an advantage of being constant-amplitude, and can be amplified with Class C amplifiers without distortion.

2.2.2 Propagation Characteristics

As in the FDMA system, the small slice of spectrum given to cellular operators makes the selection of antennas, the spectral efficiency and the capacity of the system very important considerations.

2.2.2.1 Typically Used Antennas

Like the FDMA systems, the antennas used in startup TDMA systems are initially omnidirectional. The 120-degree sectorized antennas replace them as the system matures, enabling increased frequency reuse and reducing the effect of cochannel interference.

2.2.2.2 Spectral Efficiency

Based on the specification, the data throughput is 270.833 kbps and the channel bandwidth is 200 kHz, making the spectral efficiency 1.354 bits/sec/Hz. The power spectral density (PSD) of a baseband GMSK signal is shown in Fig. 2.9. As shown in the figure, if an adjacent channel signal is present in a receiver, the overlap in the spectrum is not negligible. For this reason, it is general practice to geographically separate adjacent channels.

2.2.2.3 Capacity

In determining capacity for the GSM system, blocking probabilities given a number of channels, or vice versa, can be found by using the Erlang-B tables. In some respects, the GSM system can be likened to an FDMA system with 1000 available duplex channels in terms of capacity. Non contiguous channels are assigned to cells to minimize the effect of sidelobe interference, but with K=7, using all of the channels could cause unacceptable interference, thus reducing the true number of channels.

2.2.3 Channel Attributes

As stated earlier, a GSM channel is 200 kHz wide and is divided into eight time slots. In each time slot, 13 kbps are reserved for speech coding, and the rest (about 20 kbps) are

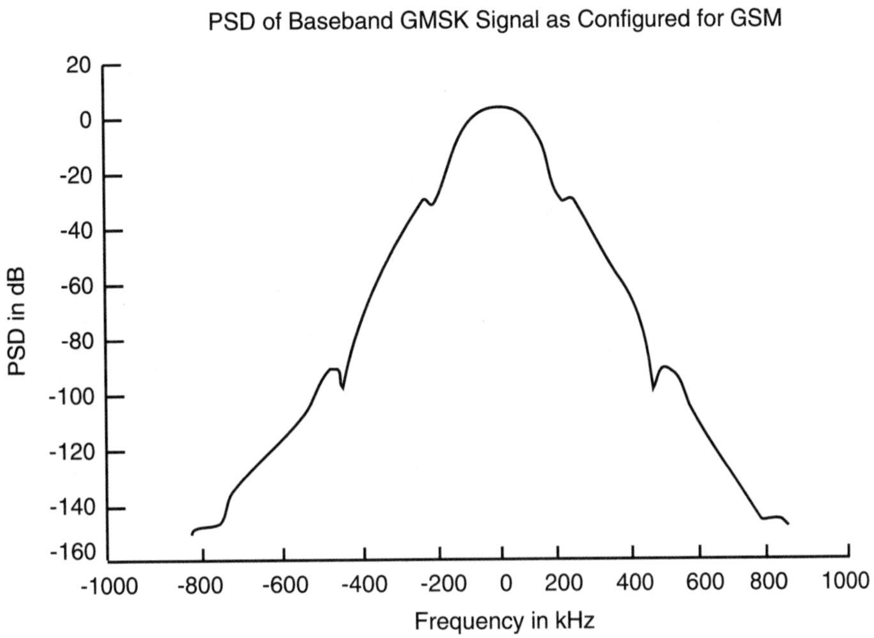

PSD of Baseband GMSK Signal as Configured for GSM

Figure 2.9 Power Spectral Density of Baseband GMSK Signal

reserved for overhead signaling. The overhead signaling is used for things like transmit power control, future hand-off monitoring and link performance.

Propagation of the TDMA signal is analogous to that of the previously discussed FDMA system. TDMA, however, is less susceptible to frequency selective fading because the information in each time slot is spread over 200 kHz, as opposed to 30 kHz in the AMPS system. This allows for a more reliable radio link.

The minimum C/I ratio for TDMA systems is 7 dB, which is based on a 90 percent probability that acceptable communications will occur if the C/I ratio is above this value. This minimum C/I ratio ensures a minimum acceptable quality of service to subscribers while minimizing cost. [5]

2.2.3.1 Measurement Reports (GSM)

Both the mobile unit and the base station need to transmit overhead data in order to maintain a good quality voice channel. This data is delivered via the slow associated control channel (SACCH), which is part of the traffic channel (TCH) 26-multiframe. Measurement reports are provided through the SAACH every 480 ms. The mobile station measures the receive level of the serving base station, the quality of the receive signal, the receive level and ID codes for up to six adjacent cells. The base station measures the receive level and signal quality of each mobile unit, the distance to the mobile unit, and the transmit power of the mobile and base station.

2.2.3.2 Power Control (GSM)

RF power control is used to minimize the possibility of cochannel interference. Power control is effected by computing the minimum required transmit power in order to maintain a good signal power level and a good quality level. Signal power level is determined by averaging the incoming signal level over a certain period of time. This receive level (in dBm) is mapped to a value between 0 and 63 with 0 equal to -111 dBm. Quality level is determined by computing the bit error rate (BER). The BER is mapped to eight levels, where 0 is the best quality (BER < 2 E-3). Once the base station determines the minimum required transmit power, it sends this information to the mobile unit via a 5 bit transmit power field in the SACCH. The transmit power of the mobile unit is variable from its maximum power to 20 mW in 2 dB steps.[6]

In the base station, transmit power control may be used but is optional. From an implementation standpoint, the transmit power must be set, based on the most faded mobile unit.

2.2.4 Call Origination Sequence

Prior to call origination, the mobile unit must make itself known to the base station. Once the unit has made itself known, the mobile unit is in the idle mode. Once in the idle mode, the mobile unit is ready to originate and terminate calls.

From this point, the description of call origination becomes heavily laden with the use of acronyms. Rather than jump into the description of all of the access channels and the framing formats, we can say that call origination in GSM follows roughly the FDMA scheme. GSM Recommendation 04.07 and [5] handle the sequence of call origination; the reader is referred to those documents for more details.

2.2.5 Call Termination Sequence

Like call origination, call termination in GSM is similar to that of FDMA. When a call comes in from the landline, the base station will page the unit and, if it is available, initiate a channel setup routine similar to the FDMA sequence.

2.2.6 Description Of Hand-Off

The hand-off process is implemented in the base station, mobile unit, and Mobile Switching Center (MSC), which is analogous to the MTSO in the AMPS system. The decision to hand off is made in the MSC, based on the measurements by the mobile unit and the base station. The measurements used by the MSC are shown in Table 2.3.

There are two types of hand-off in the GSM system, intercell and intracell hand-off. Intercell hand-off normally occurs when the current serving cell shows a low receive level and a low receive quality and a surrounding cell has a better receive level, or when a surrounding cell allows communication with a lower transmit power level. Intracell hand-off occurs when the receive quality level is low and the receive level from the serving cell is

Table 2.3 Base Station and Mobile Station Measures Used in Determining Hand-off.

Base Station	Mobile Station
Uplink performance	Downlink performance
Signal strength of interference on idle channels	Signal strengths received from neighboring cells

high. This shows a degradation due to interference well within its cell. The intracell hand-off can hand off to a different time slot on the same channel or to a different channel altogether. [5]

During normal operation, the mobile station must identify up to six neighboring cells and report their signal strengths to its serving cell. Because of frequency reuse, the carrier frequency of an adjacent cell does not necessarily identify it uniquely. The mobile station must, therefore, demodulate and decode the surrounding cells' carriers and obtain the base station identification code (BSIC). These codes, along with their signal strengths, are then transmitted back to the base station for processing by the MSC. [5]

All algorithms for hand-off are based on the following generic procedure [6]:

1. An average of the received power level (dBm) and signal quality (BER) is computed over a period of time defined in the GSM specification.

2. The average values are compared to thresholds. A hand-off algorithm (several of which are described later) will determine whether hand-off should be performed. The algorithms base their decisions for intercell hand-off on signal quality, distance and power budget. For intracell hand-off, decisions are made based on signal quality, as low signal quality is usually caused by in-band interferers.

3. Hand-off requests are then processed by the MSC for intercell hand-offs or by the base station controller (BSC) for intracell hand-offs.

2.2.6.1 Hand-Off Algorithms

Three hand-off algorithms will be described:

1. The algorithm described by [5] : GSM 05.08.
2. Fuzzy controlled hand-off [7].
3. Pattern recognition by hidden Markov models [8].

2.2.6.1.1 GSM 05.08 ALGORITHM: As stated earlier, all of these algorithms for making hand-off decisions follow the basic routine of taking measurements, making a comparison (decision), and possibly handing off. Once the measurements are made, threshold comparison begins. Table 2.4 lists the thresholds. For each threshold, a confidence factor is added by making the decision based on the last N averages exceeding the threshold P times; for example, if P=10 and N=12, then the mobile unit's transmit power will be increased if the base station reports a receive level less than the L_RXLEV_UL_P threshold for 10 out of the last 12 measurement periods.

Based on the comparison of the measured values with the thresholds in Table 2.4, the base station controller will then decide, based on an algorithm, whether hand-off will occur. The thresholds to control power level are included in the hand-off algorithm to reduce the possibility of an unnecessary hand-off. For example, if the power level was just reduced and a hand-off due to receive power level is issued, the hand-off should be blocked until the power level is increased. The base station controller decides which is the best cell to hand off to and sends a hand-off-required message to the MSC. The identified list of base stations by the mobile unit, the controller prioritizes in order from best to worst choice. The best choice in a region where two base stations appear to be the same is the one with the best positive trend. For example, if the mobile unit is in the fringe areas of the serving cell and is equidistant from two adjacent cells, then the cell chosen would be the one where the received power levels have been increasing.

A hand-off is considered imperative if the received power level is below the thresholds despite power control, the quality of the link is below the threshold while the quality level is at the threshold, or the distance of the mobile unit from the base station exceeds the MAX_MS_RANGE. In the MSC, hand-off is prioritized in the event that there is more than one request for hand-off. The order by decreasing priority is:

- Receive quality level
- Receive power level
- Distance
- Power budget

Basically, the hand-off requests due to quality level take the highest priority.

Table 2.4 Threshold Variables and Descriptions

L_RXLEV_XX_P	Lower-receive, power level threshold; increase transmit power.
U_RXLEV_XX_P	Upper-receive, power level threshold; decrease transmit power.
L_RXQUAL_XX_P	Lower receive, quality level threshold; increase transmit power.
U_RXQUAL_XX_P	Upper receive, quality level threshold; decrease transmit power.
L_RXLEV_XX_H	Lower receive, power level threshold; hand-off might be required.
L_RXQUAL_XX_H	Lower receive, quality level threshold; hand-off might be required.
RXLEV_XX_IH	Upper receive, power level threshold. If L_RXQUAL_XX_H is exceeded, intracell hand-off might be required.
MAX_MS_RANGE	Maximum distance threshold; hand-off.
HO_MARGIN(n)	Compared against power budget for each neighboring cell to ensure the mobile station is always connected to the cell with the least path loss.

Note: XX = UL for uplink, DL for downlink.

In addition to link parameters, the MSC takes into account the loading on the system and can decide to hand off calls in order to relieve congestion. In the event of congestion, the MSC may also employ call queuing. When a hand-off request is placed in the queue, it has priority in its best cell of choice over new calls coming into that cell.

2.2.6.1.2 Fuzzy Logic Hand-Off: In the ideal modeling of a cellular system, the cells are shaped like hexagons and each cell has a distinct boundary. In reality, there are no real boundaries between cells, i.e, that is, the boundaries are fuzzy. Since the boundaries are fuzzy, then why not make the hand-off between cells based on fuzzy logic? This topic is discussed in by Junius, [7].

The GSM standard defines specific numbers for receive signal level and quality. The receive quality has already been made fuzzy in the respect that there are only eight levels of quality that the channel measures for the BER. Junius [7] then fuzzifies these discrete values to fall into overlapping categories of very low, low, medium, acceptable, good and very good. Once the fuzzification is finished, the defuzzification and rule base (algorithm) is used. Past values for the quality and level of the signal are stored to evaluate trends. The rule base given in [7] is not complete and therefore is not treated here. In determining whether a hand-off should occur, the rule base assigns a hand-off certainty between 0 and 1 for every comparison. An inference engine generates a simple yes or no for its output from the rule base.

The rule base can be modified easily to optimize load balance for high-congestion areas in a manner that makes it easier for the engineer to modify a parameter and have a good idea what the result of the modification will be.

2.2.6.1.3 Pattern Recognition For Supporting Hand-Off Decisions: Pattern recognition as described by Kenneman [8] is another viable method of making a hand-off decision. The method uses hidden Markov models to determine the location of a mobile unit from the standard GSM measurement data. In urban environments, the pattern recognition method is reliable for a limited number of locations in a cell. For example, at the fringe of a cell where only a few streets are crossing the boundary of a cell, this system would be able to recognize characteristic patterns about the measurements by the mobile unit and initiate a hand-off. A pattern recognition system such as the one described in [8] can only be used to augment a true hand-off algorithm, such as a fuzzy logic algorithm, or the algorithm described by GSM 05.08.

2.2.7 Complexity And Cost Of System (GSM)

TDMA systems employ advanced measurement techniques for determining link quality and for determining the best cell for hand-off. This creates a higher startup cost in the MSC and BSC, i.e., more expensive equipment and computers for processing. The mobile unit, however, can be cheaper and lighter because of high-speed digital electronics and the TDMA format. The GSM system for TDMA is already in use in Europe, so the research-and-development is already being recouped. If GSM were to be employed in the U.S., slight redesigns of equipment would be required in order to make the European equipment compatible with the U.S. cellular system.

2.3 CDMA STRUCTURE AS PROPOSED BY QUALCOMM

Code Division Multiple Access, or CDMA, derives its name from the fact that many users share the same bandwidth by each modulating their signals onto one of a set of orthogonal codes. As is the case for all of the structures discussed in this paper, FDMA also plays a role in defining the system. IS-95, the CDMA standard as proposed by Qualcomm, has recently been accepted by the Telecommunications Industry Association (TIA) for use in North America.

2.3.1 System Description

CDMA was envisioned by Qualcomm as a cellular system to replace the current AMPS. The engineers at Qualcomm knew that to gain a new frequency allocation from the FCC would be extremely difficult, so they decided to base their system upon the cellular telephone allocation that already existed. The allocated spectrum is divided into a number of 1.25-MHz channels, each of which is either used for forward or reverse transmission. Each channel is potentially shared by a number of users that all use a different code to modulate data in a spread spectrum transmission. For each channel, there are 64 orthogonal codes, some of which have special functions (and are therefore not used as voice channels). Qualcomm has envisioned CDMA gradually replacing the AMPS system by beginning with a small portion of the cellular spectrum (1.25 MHz has already been reserved by Qualcomm and used for preliminary testing) and slowly taking over additional bandwidth from the AMPS system. Qualcomm is also not the only company interested in CDMA. Nynex has reserved 5 MHz of its spectrum for testing in and growth into CDMA. [4] Table 2.5 summarizes CDMA channelization functions.

Power control, discussed in greater detail later, is extremely important. Power in this system is controlled by using both open and closed loop methods.

Since the interference in a channel is about 50 percent due to in-cell interferers and 50 percent due to interferers external to the cell, it is important to try to minimize transmission power. To this end, CDMA systems use two primary techniques, voice compression and variable data rate. Each link has a maximum capacity of 9600 baud, which includes both data and overhead. Voice is digitized and encoded to 8550 bits per second by use of a vocoder—the other 1050 bits per second are overhead. When fewer bits are required to move the voice across the channel, such as during pauses in conversation, the transmitter has the option of reducing the data rate by factors of 2 to any of the following rates: 4800 baud, 2400 baud or 1200 baud.

2.3.2 Propagation Characteristics

Propagation in a CDMA system is substantially different than propagation in the other systems discussed herein. Although the data rate for a channel is a mere 9600 baud, it is spread across a 1.25 MHz region. Fading in the cellular frequencies, when considered in the 1.25 MHz bandwidth, tends to behave as a notch filter instead of outright destroying the channel. The large amount of error detection and correction built into the signal is usu-

Table 2.5 CDMA Channelization Functions

Parameter	Function	Notes
Frequency	Divides the spectrum into several 1.25-MHz frequency allocations.	Forward and reverse links are separated by 45 MHz.
Walsh Codes	Separates forward link users of the same cell.	Assigned by cell site. Walsh code 0 is always the pilot channel. Walsh code 32 is always the sync channel.
Long Code	Separates reverse link users of the same cell.	Depends on time and user ID. The long code is composed of a 43-bit long PRBS* generator and a user-specific mask.
Short Codes, also called the I and Q spreading sequences.	Separates cell sites or sectors of cells.	The I and Q codes are different, but are based on 15-bit-long PRBS generators. Both codes repeat at 26.667 ms intervals. Base stations are differentiated by time offsets of the short sequences.

*PRBS is defined as a pseudo-random bit sequence.
Source: Reprinted from [4].

ally able to overcome that sort of interference, especially if bit interleaving is used as expected.

Multipath signals are also used to the advantage of the receiver. Rather than allowing multipath signals to degrade or destroy the signal, the receiver uses a RAKE technology. The name RAKE is derived from the block diagram of the receiver, which includes a delay line with multiple taps. Because of time diversity, the different delayed signals are weighted and combined to give an optimal signal reconstruction.

The RAKE receiver, as mentioned above, is a correlation receiver. It works by using multiple copies of the received signal, delaying them, scaling them and then adding them together. If the delays and the scaling are correct (the result of much processing), then the signal quality will show significant improvement. Figure 2.10 is a sample block diagram of a RAKE receiver.

CDMA works because all of the transmitters operate at roughly the same power. Reference [4] compares the receiver's function to listening to a single person talking to you in a room full of conversations in other languages. The lone person speaking an understood language would be easy to discern—unless he happened to be whispering while all of the others are shouting. Tight power control ensures that multiple separate conversation participants will all be able to hear their respective speakers.

Two forms of power control are used, open loop and closed loop. The concept of open loop is based on the assumption that the loss on the forward path and the loss on the reverse path are similar. The mobile radio uses the received power (from the cell site) as a

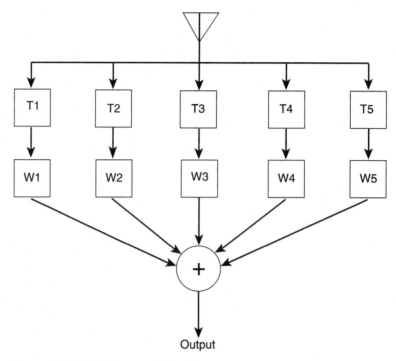

Figure 2.10 Block Diagram of RAKE Receiver
Source: Reprinted from [4].

reference: The weaker the received power, the further away the cell site transmitter is assumed to be. The mobile station's transmitted power level is set based upon assumptions about the distance to the base station. The sum of the transmit and receive powers measured in dBm (at the mobile station) is a constant -73 dBm. Reference [4] gives the example that if the received power is -85 dBm, the transmitted power will be +12 dBm.

The open loop power control is further refined by the use of closed loop power control, which forces deviation from the open loop setting. Closed loop power control is effected from the base station, which sends power control bits to the mobile station every 1.25 milliseconds. Power is controllable in increments of 1 dB.

The net effect of the power control used in CDMA is that the average transmitted power is significantly lower than the power transmitted in the AMPS system. Not only does this positively affect the amount of noise (meaning, traditional white noise plus ignition noise plus undesired signals), but it also allows longer battery life because the amplifier within the unit does not often have to operate at full power. [4]

2.3.2.1 Typically Used Antennas

CDMA uses the entire cellular frequency allocation in every cell, including adjacent ones. Although it is necessary for cells to overlap, it is desirable to keep the area of overlap to a minimum. To that end, directional (sectorized) antennas could prove useful. Omnidirectional antennas would probably still be used on base stations.

2.3.2.2 Spectral Efficiency

Spectral efficiency in CDMA is not measurable by analytic methods. Since CDMA is a new technology, the practical maximum number of channels has not been determined either by implementation or through theoretical calculation. Spectral efficiency will remain unknown until the maximum number of orthogonal codes that can be packed into a channel with an acceptable BER is determined. As digital signal processing (DSP) sophistication increases, the spectral efficiency should continue to improve as the number of links which may be packed into a channel grows.

Based on the CDMA specification, the maximum spectral efficiency can be calculated assuming 100% link usage. With 64 orthogonal codes and a link data rate of 19.2 kbps after forward error correction encoding, the spectral efficiency is 0.999 bits/Hz.

2.3.2.3 Capacity

Increased capacity is the driving force behind CDMA. CDMA's proponents believe that the spectrum that is assigned to cellular radio is all they are going to get—and they intend to make the most of it. CDMA can pack more links into a limited bandwidth than any of the other schemes examined herein, and probably with a better chance of reception. CDMA is also the only system which is capable of 100 percent frequency reuse, meaning that every bit of the spectrum is available in every cell. There are, however, other factors that tend to limit the system to a less than complete reuse of all frequencies in each cell, such as the soft hand-off feature, which is discussed in a later section. Other factors determining capacity for this system include:

Loading
- The number of calls placed by the system's subscribers
- The way the calls placed are distributed in time
- The average time per call
- The amount of overhead required

System Layout
- The soft hand-off algorithm
- The size of the cells (a major concern for this system!)

The charts and tables that developed for capacity calculations were created for use with the AMPS system, in which the number of channels required corresponds to a number of frequencies that are needed. In CDMA, however, the number of channels given in an Erlang B table would correspond to the number of 1.25 MHz channels times the number of Walsh codes available for voice links. Some experts doubt that 100 percent reuse in every cell is achievable, but if it were, it would allow each channel to be used in each cell, a sevenfold increase in bandwidth. Substantial work remains to be done in calculating the true capacity of CDMA, but it seems certain that it should far exceed that of the AMPS, NADC and TDMA systems. Analytically, calculating the capacity of this sort of system should prove almost impossible, but simulation should provide some insight. Some excellent tools for communications system modeling and simulation have been developed.

2.3.3 Channel Attributes

In the AMPS system, the maximum allowed transmitted power of the land station is 100 watts at a HAAT of 500 feet. A typical implementation, however, has a transmitter power of 20 watts and a HAAT of 100 feet. [2] The FCC will probably not raise its limits if a CDMA signaling format is used instead of FDMA but will probably not lower it, either. It therefore seems safe to assume that the transmitter powers and regulations should be the same as in the current AMPS system.

The transmit powers of the mobile radios will be tightly controlled, using open loop power control supplemented by closed loop power control.

Depending upon the type of antenna used, the gain can vary from 8 or 9 dB, depending upon whether an omnidirectional or a sectorized antenna is used. It is envisioned that the mobile units will use an omnidirectional antenna, whereas the cell sites will probably make use of either omnidirectional or sectored antennas.

The path loss with CDMA will be less subject to the influences of foliage than the path loss of other signaling formats because of the wider bandwidth of the CDMA channels. Foliage loss will look like a notch filter rather than a broadband attenuator to a signal of the width of a CDMA filter. The path loss is expected to occur much the same as the path loss for the AMPS system, with one important exception: Multipathing will not be as much of a detriment to the CDMA system as it is to the others because of the design of the RAKE receiver, which is a sort of correlation processor. It would probably be wise to treat the path losses the same as the path losses for the AMPS system, but to add a subjective receiver gain to the calculation of the link budget. The subjective gain would probably be different for areas of different terrain and building density and would have to be derived by experimentation. The block diagram for a RAKE receiver is depicted in Figure 2.10 on page 38.

One of the major advantages of CDMA is its insensitivity to noise. Because of the modulation scheme used, the required C/I ratio for a CDMA system is -18 dB after the coding gain of 21 dB is factored into the equation. [4] This means that the interfering signal power can be 64 times the carrier power and still be able to adequately discern the carrier signal.

All of these building blocks can be used to assemble a link budget. The lack of firm data for a number of the blocks precludes an actual calculation, but the above definition of the blocks should give a good idea of how to construct the link budget when the appropriate tests, measurements and designs have been performed.

2.3.4 Call Origination Sequence

When a CDMA mobile radio turns on, it is already aware of the service's frequency allocation in the local area. It will tune to the appropriate frequency and search for pilot signals, probably finding several. These pilot signals, one per base station, are distinguished from one another by use of a time offset. The mobile radio will pick the strongest reference and establish a frequency reference and a time reference from it by demodulating Walsh number 32 (one of the channel assignments), which is always used for the synchronization function. The synch channel message contains the future contents of the 42-bit long code shift register (see Table 2.5 on page 37). Three hundred twenty milliseconds (or a time of 320 milliseconds) have been allowed for reception of the long code and loading it into the appropriate register for use. [4]

Once the mobile radio is in synch, a user makes a call by entering the telephone number of the individual to be contacted and presses the send button. The mobile radio uses a page to attempt contact with the base station. A long code mask is used, based upon the cell site parameters. Therefore, if two mobile radios attempt contact with the cell site at the same time, a collision will occur. The mobile radio will know that a collision has occurred if the cell site fails to acknowledge contact with the mobile radio. In the event of a collision, each mobile radio waits a random amount of time, then tries again (classic CSMA/CD signaling). After contact is made via the paging channel, the base station will assign a traffic channel to the mobile radio by giving it a Walsh number to use. At this point, the mobile station changes its long code mask to one based on its serial number, transmits and receives in its assigned Walsh number, and may begin verbal communication. The transmit and receive channels are located 45 MHz apart and most likely will use the same Walsh number.

2.3.5 Call Termination Sequence

Information was not available on how a call is terminated. Termination could occur in one of three ways or perhaps in combinations of them.

Embedded notification: Termination by a mobile radio could be signaled by having a small digital code sent to the MTSO at the end of a conversation. This code could be repeated by the mobile radio on the transmit channel until it is acknowledged by the MTSO on the receive channel. Upon reception of the close channel message by the mobile radio, the mobile radio could cease transmission. After losing the mobile radio, the cell site could then cease transmission.

Paging channel: An additional channel could be allocated for transmission of a "close channel" notification from the mobile radio to the cell site. This has two disadvantages: use of an extra channel for overhead and the possibility of collisions.

Timeout: This option would probably be incorporated into any system. Basically, if the cell site (and hence the MTSO) does not receive transmission on an assigned channel within a certain amount of time, the link would be assumed to have been dropped and would be returned to the pool of channels, which could be assigned upon demand.

Timeout would probably be used on any system to ensure that channels would be returned to the system to be reassigned in the event of a problem. Embedded notification is probably the most elegant solution, requiring no extra bandwidth but allowing the mobile radio to completely communicate its needs to the cell site and ultimately the MTSO.

2.3.6 Description Of Hand-Off

CDMA systems support the use of soft hand-off technology. Soft hand-offs provide spatial diversity and better cell-fringe performance, but at the expense of decreasing the channel capacity of the system and increasing the amount of noise in the system, since the number of users within the 1.25-MHz bandwidth increases and the soft hand-offs look like noise to other users.

Qualcomm's system allows three links to be simultaneously operated, with a fourth link scanning in search of better channels. The number of links used by a system may be determined by first defining a signal level above which no extra help is needed. If the strongest pilot signal exceeds this level, then no "soft" links are necessary. However, if the pilot falls below the threshold, the processor will attempt to set up additional channels that operate simultaneously with the original channel and carry exactly the same information. Additional channels will not be set up if their levels are below a second threshold. A soft hand-off is in operation at any time in which two or more channels are simultaneously in operation with the same mobile radio. It is important to note that two links will not be set up with the same base station.

Bowman [10] has conducted extensive simulation of the soft hand-off process and shown the following results.

- A small percentage of soft hand-offs will occur if the thresholds are separated by 1 dB.
- A small percentage of mobile users will use three links if the thresholds are separated by 2 dB.
- When the thresholds are separated by 5 dB, a fourth pilot tone sometimes exceeds the lower threshold; thus, four base stations are viable communications link candidates.
- When the thresholds are separated by 10 dB, the number of links in action is roughly double the number of calls in progress.

All of these generalizations assume a uniform distribution of callers.

According to Bowman [10], the spatial diversity compensates for some of the loss of capacity, with the net result that for a 10 dB threshold difference, the reduction in capacity is 8 percent. Bowman points out that the 8 percent capacity reduction is a small price to pay for a 7 to 10 dB improvement in system performance, a point which is certainly worth mentioning. The beauty of this system lies in the capacity reduction and the fact that system performance improvement can be dynamically assigned. That is, the threshold difference can be changed—a smaller difference in peak times and a larger difference during times of less usage. The number of simultaneous links a mobile radio is allowed could also be changed from three to two in times of heavy loading. With this system, one can utilize the extra capacity when necessary and utilize maximum space diversity when loading allows.

2.3.7 Complexity And Cost Of System

There is no denying that this system is the most complex of those discussed in this chapter. This means that the processing software and hardware will be the most expensive. The cost of the mobile station will also be dramatically higher because of the sophisticated nature of the spatial and time diversity that must be incorporated. This translates to a higher cost to both the user and the provider.

2.4 CONCLUSIONS (FDMA, TDMA AND CDMA)

Multiple access techniques have been described and evaluated for their ability to serve subscribers on the basis of cost, efficiency and capacity. Each can serve the customers well, depending upon the size of the customer base. AMPS is the first high-capacity system to have been implemented in this country. It is still an excellent choice for the implementation of a cellular system in a market with a small number of subscribers. The system operator's cost of startup is low, as is the cost of the mobile radio for the user. As the system grows, the AMPS will eventually become insufficient. At that time, it will be necessary to transition to another system with higher capacity. Candidates are the NADC and the CDMA systems. NADC has the advantage of using the same 30-kHz channels as the AMPS, making rechannelization unnecessary while still providing roughly a threefold increase in capacity. CDMA may be inserted into a portion of the spectrum and used side-by-side with the AMPS. This ability to transition into the newer technologies is extremely important because nothing makes a customer angrier than having a system suddenly not support a communications device which he recently purchased. It takes a long time to regain customer confidence. For this reason, GSM will probably not be able to break into the U.S. market without a new spectrum allocation.

When the capacities of each system are compared, the average bandwidth per user can be computed for each system to obtain a relative spectrum use baseline. NADC has the highest bandwidth per user efficiency (10 kHz/user), followed by CDMA and GSM at close to 25 kHz/user, and, lastly, by AMPS with 30 kHz/user. The user bandwidth efficiency for CDMA was calculated using an assumed 50 traffic links per 1.25-MHz channel.

Link performance is a function of the required C/I ratio. As the C/I ratio increases, the noise performance and baseband S/N ratio improve. The dominating noise source in the cellular environment is cochannel interference rather than the thermal noise floor. With a lower required C/I ratio, a system will be less susceptible to interference from other cellular users, creating a more robust system. CDMA has the clear advantage because of the spread spectrum technique used. CDMA is also the most complex and costly system to design and construct.

Cost is a driving factor in determining whether a particular standard will be accepted by the marketplace. If a system costs too much to build and operate, it will not survive. In light of this and because GSM and NADC have already been implemented, CDMA may take a while to catch on. Advances in computer technology will make CDMA more attractive in the future for both the increased capacity potential and the advanced features CDMA offers. AMPS is by far the cheapest system to construct, build, operate and use. It is limited by the number of users it can handle, but its cost will make this system attractive for many years to come. The complexity of a system is closely tied to cost. As a particular standard matures, the system operators begin to see a return on their initial investment and the cost to the subscriber decreases. The initial cost for new mobile units also decreases as technology advances make manufacture of phones more affordable.

Each standard has its own strengths and weaknesses. Cellular operators will select a standard to use based upon their subscribers' numbers and requirements. In the end, the user will choose which system he wants to use, based upon what services are offered and how much they cost. The casual user may pick the analog AMPS system for its low cost as long as it is offered, but the business user who depends on high-quality mobile phone service may want to use one of the digital schemes described.

REFERENCES

[1] Mehrotra, Asha. *Cellular Radio Analog and Digital Systems*. Artech House, 1994.

[2] Lee, William C. Y. *Mobile Cellular Telecommunications Systems*, New York: McGraw-Hill, 1989.

[3] Couch, Leon W. *Digital and Analog Communication Systems*, 4th Ed., (New York: Macmillan Publishing, 1993).

[4] Whipple, David P. "The CDMA Standard," *Applied Microwave & Wireless*, Spring, 1994: pp. 24-37.

[5] Mouly, Michel, and Marie Bernadette Pautet. *The GSM System for Mobile Communications*, published by authors, Palaiseau, France, 1992.

[6] ETSI/TC GSM Recommendation 5.08, Version 3.7.0. "Radio Subsystem Link Control," January, 1991.

[7] Calhoun, George. *Digital Cellular Radio*, Norwood, Mass: Artech House, Inc. 1988.

[8] Junius, Martin. "Intelligent GSM Radio Resource Management Using Neuro-Fuzzy Technology," no date given (paper).

[9] Kennemann, Olrik. "Pattern Recognition by Hidden Markov Models for Supporting Handover Decisions in the GSM System," no date given (paper).

[10] Bowman, Martin D. *Analysis of Hand-Off in Code Division Multiple Access Cellular Mobile Systems*, unpublished, November, 1993.

Chapter 3

Unlicensed Personal Communications Services (UPCS) Devices

Unlicensed personal communication services (UPCS) devices are a new class of personal communication services (PCS) devices that the Federal Communication Commission (FCC) is planning to sanction soon. Unlike most types of devices using electromagnetic (EM) spectrum, UPCS devices will not be licensed by the FCC, although they will be type approved and regulated by the FCC. The FCC rules concerning the operation of UPCS devices are very strict about certain characteristics, such as emission bandwidth, peak output power, and channel access procedures, but are almost silent about modulation characteristics. Also, unlike most classes of equipment regulated by the FCC, these devices are restricted to peak output power levels of less than 317 mW.

The intent of the FCC and the workers who conceived of UPCS devices was that they would be used in small, well-defined areas, such as a single building or a small campus.

This chapter provides an overview of the characteristics of these devices as described by the proposed FCC rules and gives some examples. The sources of the data in this chapter are the FCC's second report and order (R&O) of November 1993 and its Memorandum Opinion and Order (MO&O) of 13 June 1993, which modified the original R&O. The final rules will be contained in the FCC rules as Part 15 Subpart D.

A unique feature of UPCS is that temporary authority to coordinate the operation of these devices will be granted by a private corporation, UTAM, Inc. Additionally, UTAM,

Inc., is responsible for coordinating the transition of private operational-fixed microwave service (OFS) devices from the UPCS band into other bands authorized by the FCC. UTAM, Inc., is composed of corporations that are allowed to develop and produce UPCS devices. Corporations not belonging to UTAM, Inc., will not be allowed by the FCC to produce UPCS equipment. UTAM, Inc., will use the proceeds from UPCS coordination fees to assist OFS users to relocate their equipment.

The FCC has designated UTAM, Inc., as the organization to coordinate the location of UPCS devices and to ensure that these devices are not transported from their authorized usage areas. UTAM, Inc., is responsible for developing rules that will assure that UPCS devices will not operate when they are transported into unauthorized areas. (These rules do not currently exist.)

3.0 UPCS DEVICE DESCRIPTION

The FCC has divided UPCS devices into two classes: asynchronous and isochronous. Asynchronous devices transmit radio frequency (RF) energy at irregular intervals and are typically elements of data networks. Asynchronous devices are allowed to operate within the frequency band of 1910–1920 MHz. Such devices are anticipated to compose wireless local area networks. Isochronous devices transmit RF at regular intervals and are usually parts of voice communications systems. Isochronous devices are allowed to operate within the frequency band of 1920–1930 MHz. These devices are anticipated to provide mobile voice services such as microcells for mobile communication systems. Although the current anticipation is that asynchronous devices will be used for data and isochronous devices will be dedicated to voice, there is no reason to suppose that system designers will not cross these boundaries when it will suit their purposes to mix voice and data or when the other class of device might better suit the designer's purposes. All transmissions must automatically cease if the device has no data to transmit or if the device suffers a failure. All transmissions must be digitally modulated.

The peak transmit power of a device shall not exceed 100 mW times the square root of the bandwidth in hertz. The bandwidth is frequency distance between the farthest points above and below the highest peak of the carrier, which are 26 dB below the level of that highest peak (dBc), as shown in Figure 3.1. In no case are these -26 dBc points allowed to lie outside of the allocated spectrum window. In addition to the peak transmit power requirement, the transmitted power spectral density may not exceed 3 mW in any 3 kHz bandwidth. If the device transmit antenna's gain exceeds 3 dB with respect to an isotropic antenna (dBi), the peak transmit power must be reduced by the amount that the antenna gain exceeds 3 dBi.

The monitoring threshold for UPCS devices is 30 dB above the thermal noise power for a bandwidth equivalent to its emission bandwidth. For devices that do not generate the allowed maximum power, this threshold may be relaxed by 1 dB for every 1 dB that the emission power is below the maximum. The receiving antenna for the device must have the same coverage area as the transmit.

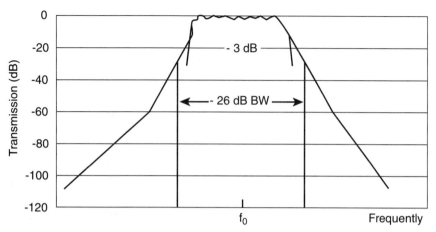

Figure 3.1 26 dB Bandwidth

3.1 ASYNCHRONOUS DEVICES

Asynchronous devices can have any bandwidth that the device designer desires between the minimum for asynchronous devices of 500 kHz and the full 10 MHz of the band.

Before an asynchronous device can commence a transmission burst, it must determine that its intended bandwidth is unoccupied. It must monitor a bandwidth equal to or greater than its intended bandwidth for a period of 50 ms or greater. If no signal is detected, the device must commence transmission within a reaction period of 50 * SQRT(1.25/emission bandwidth in MHz) ms or 50 ms, whichever is greater.

Once a device determines that it can commence transmission, it need not monitor the spectrum window again until it and its cooperating devices cease transmission. However, a group of cooperating devices may not continue to transmit for more than 10 ms. An interburst gap between cooperating devices may not exceed 25 ms. Individual unit intraburst transmission shall be separated by a uniform random-duration interval evenly distributed between 50 ms and 375 ms.

If the monitor antenna determines that the desired spectrum is occupied, the device may change to another frequency and monitor or conduct the following process:

1. Wait a deference period randomly chosen from a uniform random distribution of 50 ms to 750 ms.
2. Repeat the access procedure. If the spectrum window is unoccupied, the device may commence transmission after the reaction time. If the spectrum window is occupied, go to step 3.
3. Each time the access attempt fails, the deference time will be doubled until an upper limit of 12 ms is reached. Twelve ms remains the difference time until the device can access the channel.

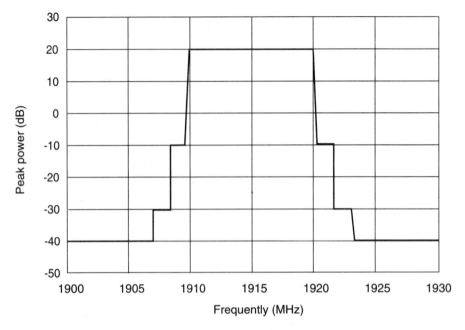

Figure 3.2 Out-Of-Band Emissions Mask For Asynchronous Devices

Each time the device and its cooperating group of devices complete a burst, the device must repeat the random access procedure before it can commence another transmission.

If a system of devices occupies less than 2.5 MHz, it must start searching for an unoccupied spectrum window within 3 MHz of either band edge. Devices of 2.5 MHz or greater will start their search at the center of the band. Devices of 1.0 MHz or less may not occupy the center half of the band if there is an adequate spectrum window in either of the outside quarter bands.

While the rules for asynchronous devices do not allow for full-duplex operation using separate frequencies for transmit and receive, they do not specifically prohibit such operation if the above rules are followed for each frequency to be used.

Out-of-channel emissions for asynchronous devices shall be less than -9.5 dBm for emissions between the channel edges and 1.25 MHz above and below the channel edges. Between 1.25 MHz and 2.5 MHz above or below the channel edge, emissions shall not exceed -29.5 dBm. The mask for the out-of-band emissions is shown in Figure 3.2.

3.2 ISOCHRONOUS DEVICES

The isochronous band is divided into eight 1.25-MHz channels. The channels may be further subdivided so long as no subchannel is less than 50 kHz in width. Cooperating isochronous devices may operate on different frequencies as long as each time and spectrum window is determined to be unoccupied prior to commencement of transmissions.

Although a system may use a frame period of any desired submultiple of 10 ms that the designer chooses, the device must monitor the spectrum and assign transmission windows assuming a 10 ms frame period. Once a device has monitored for at least 10-ms and determined that the desired time and spectrum windows are available, it and its cooperating devices may commence transmissions. The group of cooperating devices may continue to occupy continuously the same time and spectrum windows for a maximum of 8 hours without reinitiating. If, however, no cooperating device acknowledges transmission within any 30-second period, the system must cease transmission and reinitiate.

If, during access procedures, a device determines that the desired time and spectrum windows are occupied, it may select another set of time and spectrum windows or wait for these windows to become available and commence using them within a randomly chosen waiting period of between 10 ms and 150 ms, commencing when the channel becomes available.

The frame period for isochronous devices is 10 ms or an integral submultiple of 10 ms. If the system uses time division to achieve a duplex connection, the frame rate frequency stability must be maintained at 50 parts per million or better. The frame jitter rate may not exceed 25 ms for any two consecutive transmissions.

Out-of-channel emissions for 1.25 MHz channels shall be less than -9.5 dBm for emissions between the channel edges and 1.25 MHz above and below the channel edges. Between 1.25 MHz and 2.5 MHz above or below the channel edge, emissions shall not exceed -29.5 dBm. Beyond 2.5 MHz above or below the channel edges, emissions shall not exceed -39.5 dBm. The mask for the out-of-band emission is shown in Figure 3.3. For systems that use a more narrow channel, the mask is:

- -40 dB with respect to the peak power allowed for the emission bandwidth in the band between one bandwidth (BW) and 2 BW from the center of the emission bands
- -50 dB with respect to the peak power allowed for the emission bandwidth in the band between 2 BW and 3 BW from the center of the emission band
- -60 dB with respect to the peak power allowed for the emission bandwidth in the band between 3 BW from the center of the emission band and the edge of the 1.25 MHz channel

3.3 UNLICENSED PCS APPLICATIONS

The future of PCS is uncertain at this time. However, the unlicensed portion of the spectrum offers great opportunity for technological exploitation.

While billions have been spent on licenses for the 1850–1990 MHz band, the unlicensed (UPCS) 1910–1930 MHz and 2.39–2.4 GHz bands offer spectrum for the entrepreneur, who does not have to raise large amounts of money. The innovative use of the unlicensed band should lead to "new" PCS services in the future. Perhaps the Dick Tracy

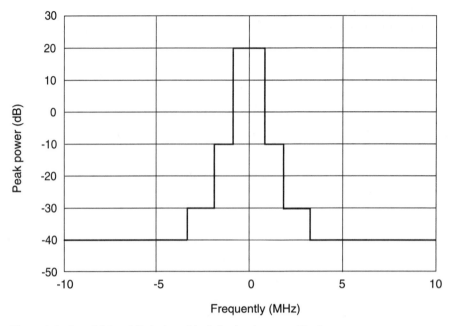

Figure 3.3 Out-Of-Band Emissions Mask for Isychronous Devices

wristwatch will become a reality within the next few years. The real challenge is to develop techniques in this 20 MHz allocation that will lead to new services.

Noncellular techniques could be developed to offer voice, message and video services within businesses. A small hand-held device could be used to confer with colleagues in an investment office, factory floor or retail store. Wouldn't it be helpful in a large retail store like Wal-Mart to have voice, video and message service between the personnel in the stock room and on the floor? It certainly would be useful for doctors to have a device to get more accurate information on their records. From a data bank they could get patient information that would be helpful in treatment. Police could have access to video and text information in carrying out their duties of surveillance and arrests.

The applications are very wide-ranging and diverse. The challenge is to design and develop PCS services that utilize a band of frequencies that do not require large capital outlays initially. The techniques for digital communications have been thoroughly researched. What is required now are developments to meet the specific needs of the business, government and home communities.

Both the asynchronous and isochronous bands could be used for special applications. Chapter 7 discusses wireless LANs and applications using UPCS. Chapter 8 illustrates a specific application for an enterprise network and sections 3.4 and 3.5 of this chapter discuss specific UPCS applications.

3.4 AN EXAMPLE OF THE USE OF THE UNLICENSED BAND (COMPRESSED VIDEO)

The unlicensed band (UPCS) provides an opportunity to develop innovative techniques for PCS applications. The 20-MHz available spectrum can be utilized for voice, data and video communications, providing data rates in the order of 40–60 Mbps/sec. One interesting application of this frequency band would be a video overlay using broadband CDMA. A field trial [1] has demonstrated that compressed video and wireline quality audio at 64 kbps could be transmitted by using the existing cellular telephone spectrum (825–894 MHz) without affecting the normal cellular service. The technology employed was broadband code division multiple access (B-CDMA). The demonstrations showed the capability of B-CDMA equipment to support fixed and mobile users at distances of three to five miles from the base station. Both voice and 64-kbps video were transmitted. The signals were successively switched to the PSTN and onward via a fiber optic network to a central command center where the video was observed by a large audience. The participants had small hand-held sets with video cameras. Since video and voice were successfully transmitted utilizing AMPS, there is every reason to believe that it could be done at the 1920–1930 MHz isochronous UPCS band.

Techniques utilizing TDMA (IS-136) and the proposed CDMA services (IS-95) have potential, since each results in a capacity increase and provides privacy, but the voice quality remains poor, data rates are low and the systems are extremely susceptible to fading.

> "Broadband-CDMA (B-CDMA) provides higher capacity than the above mentioned technologies, increased resistance to fading, privacy, wire-line voice quality and data rate on demands up to ISDN rates, thereby permitting *multimedia* communications. Further, B-CDMA can overlay and share the spectrum with existing AMPS or TDMA cellular users." [1]

The concept for the 1920–MHz band would be similar, except that the transceiver would be multimode, that is, it would be usable for the 825–894 MHz band or the 1920–1930 MHz band. Utilizing the latter band would essentially provide for a private network for indoor or outdoor use. Doctors could use this network to perform diagnostics and to recommend treatment for emergencies. Fire and rescue units could use it for on-scene help. Retailers could use it for inventory control. The military could use it for logistic or combat command and control. Construction personnel could use it for construction coordination.

The use of the 1920–1930 MHz band should provide less path loss, both indoor and outdoor, less interference because there will be fewer users on this band and a "unique" business opportunity because the band is unlicensed.

Figure 3.4 shows a configuration for this mutlifunctional system.

UPCS devices in the 1910–1920 MHz and 2.39–2.4 GHz bands are asynchronous devices that can be utilized in the LAN/WAN network environment. Application in these frequency ranges are discussed in Chapters 7 and 8. An additional example for asynchronous UPCS is given below.

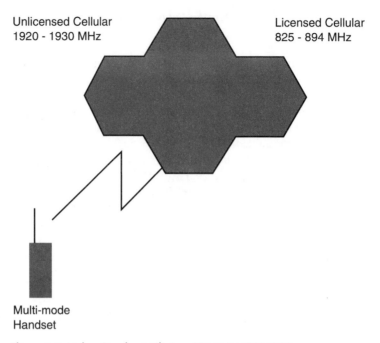

Unlicensed Cellular
1920 - 1930 MHz

Licensed Cellular
825 - 894 MHz

Multi-mode
Handset

Figure 3.4 Video Overlay Utilizing CDMA/AMPS/UPCS

3.5 LANS USING THE UPCS BAND

Aksriodi [2] writes:

"The UPCS band which is segmented into a 1.910–1.92 GHz and a 2.390–2.4 GHz subband for asynchronous applications such as wireless LANs and data provides the prospects for higher data rate and new opportunities for emerging data applications since the rules for these bands do not impose bandwidth-inefficient techniques like spread spectrum modulation. In the UPCS spectrum, for example, a device operating in the asynchronous subband must follow a special Listen-Before-Talk (LBT) etiquette, which is designed to allow multiple systems to coexist in the same vicinity. In this etiquette, devices must listen for 50 microseconds of silence before they can transmit. After devices complete transmission, they must wait for a randomly chosen time between 50 and 750 microseconds before they attempt to transmit again. Each time that the channel seems to be busy, the devices must double the maximum extent of the deferment time or until a maximum extent of the deferment time of 12 ms is reached. Further, no transmission can exceed 10 ms in length. There are also criteria governing power control, transmission bandwidth, digital modulation techniques, and spectral efficiency." [2]

Using the asynchronous UPCS band, LAN access devices could be developed for mobile e-mail and data services. Figure 3.5 indicates a possible application for wireless

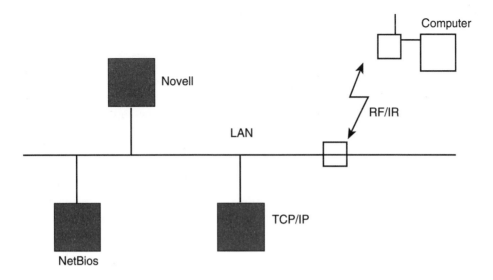

Figure 3.5 LAN Access Device

LAN access. The hand-held access device can use either the RF or IR frequencies for transmission. Although the IR devices would have limited range due to walls and other obstructions, they do provide an interesting alternative for high data rate in room coverage. IR systems are not bandwidth limited by FCC regulations and in fact appear to be capable of transmission speeds of 50 mb/s or more. [2]

In this application, the hand-held device would utilize either the RF or IR frequencies and would have a transceiver connected to the LAN. Remote access would be provided to the user within the building or in the immediate vicinity. A laptop computer could be connected to the remote device. This would enable e-mail access from conference rooms, retail or factory floors, baggage handling, etc. The remote device would utilize the UPCS asynchronous protocol.

3.6 CONCLUSIONS

This chapter discussed the requirements for a new class of personal communication devices, the unlicensed personal communication services devices. The initial intended uses for the two subclasses is that asynchronous devices will carry data and isochronous devices will carry voice, but there are no actual restrictions about device usage. The extremely low transmitted power restrictions are intended to restrict the use of systems of these devices to very small areas, such as a single building or a small campus.

REFERENCES

[1] Grieco, D.M., and W.H. Biederman, "Video Overlay on Cellular Using Broadband - CDMA." Proceedings, 1995 IEEE Long Island section, Wireless Communication System Symposium, November, 1995.

[2] Aksriodi, H., "Wireless and Mobile Networking in the Enterprise Environment." Proceedings, 1995 IEEE Long Island section, Wireless Communication System Symposium, November, 1995.

Chapter 4

Spectral Efficiency

The efficient use of the spectrum is the most important problem in mobile communications. [1] The market for cellular radio services is expected to increase dramatically this decade: Service may be demanded by 50 percent of the population. [2] The fulfillment of this demand is beyond what can be accomplished with the presently used analog cellular system. Digital technology, modulation and multiple access techniques are being developed to improve spectrum utilization. Mehrotra [1] talks about ways to improve the spectral efficiency of FM. These methods and others are discussed in this chapter to emphasize the need to improve spectrum efficiency. These methods employ use of smaller channel bandwidths, which necessitate large C/I ratios for the voice quality and they reuse distance. Some techniques improve analog FM cellular by only 60 percent, compared to digital and multiple access techniques which can improve efficiency by multiple factors. Similar techniques can be used for other access techniques.

In addition to a discussion and general analysis of spectral efficiency of multiple access techniques, a hybrid scheme of TDMA and CDMA is examined for deployment and spectral efficiency.

4.0 DEFINITION OF SPECTRAL EFFICIENCY

Two terms are used to evaluate cellular systems: channel efficiency and spectral efficiency. In conventional communications systems, where the frequency spectrum is not reused, channel efficiency equals spectral efficiency. In cellular systems, the allocated spectrum is reused. The spectral efficiency can be defined as the maximum number of calls that can be served in a given frequency spectrum:

1. voice channels/MHz/area
2. erlangs/MHz/area

Definition 1 is called the modulation efficiency and is defined:

$$(\eta_m = 1/(B_c * N * A) = \frac{B_t / B_c}{B_t(NA)} \tag{4.1}$$

Definition 2 is called the load per megahertz per area and is defined:

$$(\eta = \text{traffic of } (B_c/B_t) \text{ channels} / (B_t * N * A) \tag{4.2}$$

The voice quality depends on the size of N, the number of cells/cluster, which is a function of C/I. Also, the spectral efficiency in definition 4.2 depends on the blocking probability. Definition 4.1 can be drawn directly from definition 4.2.

Lee [2] defines a quantity called radio capacity and is derived in the appendix to this chapter:

$$m = B_t/B_c \ \sqrt{2/3(C/I)} \tag{4.3}$$

where m = number of channels per cell or "radio capacity"

B_t = total available bandwidth

B_c = voice channel bandwidth

C/I = cochannel interference

A = area of cell in Km^2

N = number of cells per cluster (cluster size)

The C/I ratio is determined through subjective listening tests. The criterion is that of 75 percent of the listeners rate the quality at good or excellent over 90 percent of the service area. For the U.S. FM system, this value is 18 dB.

In the appendix to this chapter, it is shown, that reducing the channel bandwidth and N increases the system capacity, or spectral efficiency. This is true on paper. In analog FM, as the channel bandwidth is reduced, noise susceptibility is increased, SNR, which in turn raises the C/I ratio for comparable voice quality. This action increases frequency reuse distance, which nullifies the improvement in capacity.

Spectral efficiency of a cellular system is a function of C/I and channel bandwidth. In communications systems without reuse, C/I does not come into play. This argument has been applied to single side band SSB systems: Because the channel bandwidth of SSB was 5 kHz as opposed to FM, which uses 30 kHz, SSB must be six times more efficient than FM. The following analysis [1,2] shows that this is not the case.

Three proposed bandwidths will be examined: 3 kHz, 5 kHz, and 7.5 kHz. The C/I ratio at RF is directly related to SNR at baseband. Given: 1) the required baseband SNR for the FM receiver with preemphasis/deemphasis and a two branch diversity combiner is 38 dB; and 2) the total bandwidth is 12.5 MHz. The SNR for the SSB system is also 38 dB. Since SSB is linear, SNR = C/I, as shown below. From Lee [2]:

$$SNR_{2br\ pre/de} = (C/I - 3) + G_{pre/de} + G_{2br} \qquad\qquad 4.4$$

where: -3 is loss in FM system without preemphasis/deemphasis and diversity

$$G_{pre/de} = (f_2/f_1)\ Z/3 \quad f_2 = 3000\ Hz \qquad\qquad 4.5$$
$$f_1 = 300\ Hz$$
$$= 15.2\ dB$$

$$G_{2br} = 8\ dB\ resulting\ from\ advantage\ of\ diversity\ and\ preemph/deemph$$

So:

$$(C/I)_{ssb} = (S/N)_{ssb} - (S/N)_{2br}FM = (18 - 3) + 15.2 + 8 = 38.2\ dB \qquad\qquad 4.6$$

This equation says that the required C/I for a SSB with voice quality comparable to that of FM is 38.2 dB, compared to FM's 18 dB. From equations derived in the appendix to this chapter, the cells/cluster, N, Reuse Distance (D) for FM and SSB are:

a) N

For FM: $\qquad N \quad = \sqrt{2/3(C/I)} \qquad\qquad 4.7$

$$\qquad\qquad\qquad = \sqrt{2/3(10^{1.8})}$$

$$N \quad = 6.5 \approx 7\ cells/cluster$$

For SSB: $\qquad N \quad = \sqrt{2/3(10^{3.8})} \qquad\qquad 4.8$

$$N \quad = 64.8 \approx 65\ cells/cluster$$

b) Reuse Distance, D.

For FM: $\qquad D_{FM} = R_{FM}\ \sqrt{3N} \qquad\qquad 4.9$

$$\qquad\qquad = R_{FM}\ \sqrt{3(7)}$$

$$D_{FM} = 4.6 R_{FM}$$

where R = radius of cell

For SSB: $D_{ssb} = R_{ssb} \sqrt{3(N)}$ 4.10

$D_{ssb} = R_{ssb} \sqrt{3(65)}$

$D_{ssb} = 14R_{ssb}$

One can see that cluster size for SSB is quite a bit larger. Next, the number of channels/cells will be calculated to see if capacity can be improved here. Given B = 3 kHz, 5 kHz, 7.5 kHz [2]

a) 3 kHz   4.11

$m = 12.5E6/(3E3*) \sqrt{2/3(10^{3.8})}$

m = 64.2 channels/cells

b) 5 kHz

m = 38.5 channels/cell

c) 7.5 kHz

m = 25.7 channels/cell

The last step is to determine the cell size for a balanced view of system capacity. The cell size is determined relative to a given cell size for FM. SSB systems are not 30 kHz, so the noise level for the three bandwidth selections will be adjusted accordingly:

a) 10Log(3K/30K) = -10 dB

b) 10Log(5K/30K) = -7.8 dB

c) 10Log(7.5K/30K) = -6.0 dB

The adjusted C/I relative to FM is:

a) 3 kHz $(C/I)_{ssb}$ $-(C/I)_{FM}$ = 38.3 - 18 - 10
= *10.2 dB higher*

Similarly for 5 kHz and 7.5 kHz:

b) 5 kHz   38.2 - 18 - 7.8 = *12.4 dB higher*

7.5 kHz 38.2 - 18 - 6 = *14.2 dB higher*

Table 4.1 Comparison of Modulation Efficiency

System	B_c kHz	N	Channel/cell	Cell Radius (km)	Cochannel Distance(km)
FM	30	7	57	10	46
SSB	3	65	64	5.5	77.6
SSB	5	65	39	4.9	69.1
SSB	7.5	65	26	4.4	62

Source: Reprinted from [4]

Let the FM cell radius be 10 Km and the 4th-power propagation loss applied:

$\qquad$ a) 3 kHz $\qquad$ -40Log(R/10) = 10.2 dB

$$R_3 = 5.5 \text{ km}$$

b) similarly, for 5 kHz and 7.5 kHz

$$R_5 = 4.9 \text{ km}$$

$$R_{7.5} = 4.4 \text{ km}$$

From equation 4.9, the cochannel distances for the examples are:

$\qquad$ FM 30 kHz: $\qquad$ 46 km
$\qquad$ SSB 3 kHz: $\qquad$ 77.6 km
$\qquad$ SSB 5 kHz: $\qquad$ 69.1 km
$\qquad$ SSB 7.5 kHz: $\qquad$ 62 km

The results of these calculations are summarized in Table 4.1.

$\qquad$ Though the SSB 3-kHz channel bandwidth is able to carry the greatest number of channels/cell and a small cell area, the 3-kHz bandwidth is not recommended for toll-quality services for cellular applications in a Rayleigh fading environment [2, 3]. An analog speech signal is toll quality when the frequency range is from 200 Hz to 3200 Hz, its SNR is greater than or equal to 30 dB, and the harmonic distortion is less than or equal to 2.3% [2, 3].

4.1 MULTIPLE ACCESS

Analog systems have limited capacity, which leads to the multiple access and digital techniques for improved capacity. Digital systems inherently provide greater efficiency because of their lower susceptibility to cochannel interference, which leads to lower frequency reuse distance. A narrower channel bandwidth in FM cellular leads to higher C/I ratio. For comparable voice quality, digital system requires a C/I of only 12 dB or 13 dB.

Lee [3] states that given a two-branch diversity scheme, the bit error rate of 10^{-3} in a slow-fading environment will require an E /I of 15 dB. Equation 4.12 shows the required C/I with this requirement in mind:

$$C/n = (E_b/I_o)(R/B) \qquad\qquad 4.12$$

where: R = bit rate

                 B = channel bandwidth

                 C/I represents carrier to noise ratio (CNR)

E_b/I_o = bit energy to noise

C/n   = (31.6) (16kb/sec/30kHz)

C/n   = 12.3 dB

From equation 4.7, the number of cells/cluster is:

$$N = \sqrt{2/3(10^{1.23})}$$

$$N = 3.4 \approx 4 \text{ cells/cluster}$$

Lee states that in this particular case, the number of available channels is 333; therefore, FSK FDMA provides 83 channels/cell.

4.1.1 TDMA—What Is It?

An alternative way to utilize the available spectrum is to let each user have access to the whole band (known as wideband TDMA) for a short time, that is, burst transmission, during which time the user's information is transmitted at a high data rate. The frequency allocation is shared among users via different time slots. Typically, the spectrum is only partially allocated, each user group having a slot allocation in a particular channel group. This is called narrowband TDMA and is shown in Figure 4.1.

One could set up slower frames, as apply to existing FDMA schemes. For example, ADC and JDC (American and Japanese) systems use three time slots per frame as opposed to the eight slots shown in Figure 4.1. Table 4.2 shows the different setups. Figure 4.2 shows how a typical frame is broken down.

Here, N time slots per frame are shown. Each user gets a slot, starting with 0. Here is a list of the parts:

- Header message: contains channel identity, bit timing recovery, carrier recovery, unique words
- C and S: control and signaling

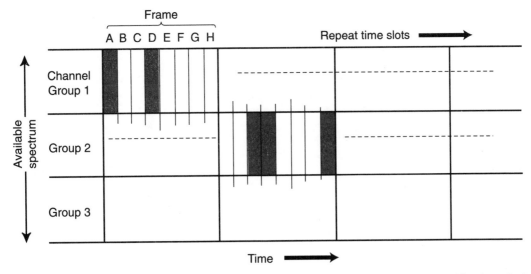

A TDMA mode of operation: each channel groups eight subscribers (A to H), who send (and receive) messages as bursts. The number of base station transmitters is equal to the number of groups.

8 subscribers A, B, C, D, ..., H per channel (group)

Figure 4.1 Narrowband TDMA
Source: Reprinted from [4].

Table 4.2 FDMA to TDMA Change and Parameters

System	TACS	ADC	GSM
Multiple access method	FDMA	TDMA	TDMA
Channels per carrier	1	3	8
Carrier spacing (kHz)	25	30	200
Number of channels	400	333	50
Frame period (ms)	No limit	40	4.6
Channel data rate (kbps)	10	48	270
Number of user channels	400	999	400

Source: Reprinted from [4].

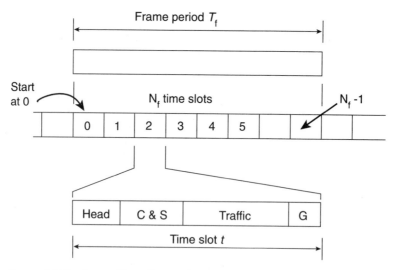

Figure 4.2 The Frame, Time Slot, and Message Relationship in TDMA
Source: Reprinted from [4].

- Traffic: the user's message
- G: guard space to allow for time/distance delays from the cell size

In TDMA systems, the channel bandwidth is an equivalent channel bandwidth, i.e., a TDMA radio bandwidth of 30 kHz is equivalent to three time slots of 10 kHz each. This means that the minimum C/I of each time slot is the same as that of the whole bandwidth. If the C/I of TDMA is 14 dB for comparable voice quality to FM, then:

$$N = \sqrt{2/3(10^{1.4})}$$

$$N = 4 \text{ cells/cluster}$$

Given a total bandwidth of 1.25 MHz, then there are 125 channels and 31 channels available per cell.

Advantages of TDMA

1. Because TDMA allows multiple circuits per carrier, money can be saved on transmitters at the cell sites.

2. Burst transmission impacts positively on the cochannel interference because at any given moment, only a part of the mobiles are transmitting, leading to better frequency reuse.

3. No duplexers are required, saving money. These can be replaced with fast switchers to turn the transmitters and receivers on and off.

4. TDMA is flexible, that is, as speech-coding algorithms improve, the TDMA channel is reconfigurable.

Disadvantages of TDMA

1. The TDMA mobile requires more-complex signal processing hardware. However, as VLSI advances, this may become a nonproblem.

2. The TDMA receiver must resynchronize on each burst. Also, because unequal propagation delays may cause one user to slip into another user's slot, TDMA needs a larger overhead than FDMA, which could be a penalty of as much as 30 percent of the total bits transmitted.

4.2 CDMA

Spread spectrum techniques (such as direct sequence and frequency hopping), which have been used in military applications because of their antijam and multipath rejection, have been proposed in the form of a Code Division Multiple Access (CDMA) to provide simultaneous digital communications. The development of CDMA is mainly for improved capacity. CDMA has virtually unlimited capacity because CDMA is only interference limited, whereas TDMA and FDMA are bandwidth limited. Reduction in interference translates to an increase in capacity. Since many users share the same channel, interference can occur if there is not enough isolation between users. [5, 6] For example, since voice signals are active inversely proportional to the time, a voice signal is silent by suppressing transmission during quiet time. This amounts to a factor of about 2.5. Another factor that can improve capacity is the 120 degree antenna/cell sectorization; this means a threefold increase. Together, CDMA gives an increase in capacity by a factor of 7.5 (3*2.5).

In earthbound mobile communications, isolation among cells gives CDMA the advantage over TDMA and FDMA. This isolation under UHF radio propagation increases by the 4th-power path loss. While FDMA and TDMA have a specific minimum frequency reuse distance, specifically a seven-cell and four-cell pattern, CDMA can reuse the same spectrum for all the cells, thereby increasing capacity over the normal frequency reuse factor. [7]

Suppose six channels are assigned per cell. In FDMA, six channels serve six calls. In TMDA, the channel bandwidth is three times wider than the FDMA channel bandwidth, so that two TDMA channel bandwidths equal six FDMA channel bandwidths. Each TDMA channel has three time slots; six time slots serve six calls. In CDMA one channel equals six FDMA channel bandwidths. CDMA can also squeeze additional code sequences into the same radio channel, whereas the other two schemes cannot. Adding more code sequences will degrade the voice quality, however.

Power control is a necessary part of CDMA. For a single cell site with power control, all reverse link signals (mobile to cell site) are received at the same power level. For

M users, each cell site processes a composite waveform containing the desired signal at power P and M-1 interfering signals, each with a power P. So, the C/I is:

$$CIRF = C/I = P/(M-1) * P \qquad\qquad 4.13$$

$$= 1/(M-1)$$

CIRF = cochannel interference reduction factor

C/I at RF is closely related to E_b/I_o at baseband.

$$C/I = (E_b/I_o)(R_b/B_c) \qquad\qquad 4.14$$

where: E_b = energy per bit

I_o = interference/Hz

R_b = bit rate

B_c = channel BW

In TDMA and FDMA there are designated time slots and channels, so $R_b = B_c$ and E_b/I_o at baseband is greater than 1. In CDMA, all N-coded sequences share the same radio channel; therefore, B_c is much greater than R_b. B_c is replaced with B_{ss}, the spread spectrum bandwidth. The interference level is always higher than the signal level; therefore, C/I is less than 1 (-x dB).

Substituting 4.13 into 4.14:   4.15

$$E_b/I_o - (C/I)/(B_c/R_b)$$

$$E_b/I_o = (B_c/R_b)/(M-1)$$

Dropping the subscripts for convenience: B/R is the processing gain of the system and is directly proportional to the number of users it can serve. E/I is the value required for a BER greater than 10^{-3}. For dual antenna diversity at the cell site, E/I = 7 dB for QPSK. Equation 4.15 does not include background and spurious noise in the spread bandwidth B_s. Including this degradation in the equation:

$$E/I = (B/R)/((M-1) + \mu/P) \qquad\qquad 4.16$$

μ/s - normalized noise (background etc.)

Therefore, the number of users becomes

$$M = 1 + ((B/R)/(E/I)) - \mu/P \qquad\qquad 4.17$$

To improve the capacity, the E/I can be increased through improved coding or modulation; however, the increased complexity and expense nullify the improvement. The other way is to reduce user interference.

As mentioned earlier, one technique is to use three 120 degree directional antennas per cell. The interference sources seen by any antenna is one-third that of an omnidirectional antenna. This reduces the denominator of equation 4.15 by a factor of three, and N is increased almost by a factor of three. So, the number of users/cell 3N equals N_3. Another technique is to suppress transmission when the voice is silent. Voice is active only 40 percent of the time. This reduces the denominator by .4, M-1 to .4(M-1). So, going back to 4.16, we have:

$$E'/I' = (B/R)/((M_3 -1).4 - \mu/P) \hspace{3cm} 4.18$$

With these factors, the capacity is roughly increased by a factor of 7.5.

An example from Mehrotra [1] will demonstrate equation 4.18. Given B = 12.5 MHz and R = 8 Kbits/sec and E/I = 7 dB, then

$$M = 1 + ((12.5E6/8E3)/5)$$

$$M = 314 \text{ channels}$$

The speech factor of 2.5 increases the capacity to 785. An equivalent FDMA with seven cell sites is 56 as shown in the appendix to this chapter. The increase in capacity is 785/56 = 14.

The above example was simplified to demonstrate the principle. In reality, it is one channel given to all users, but several channels sharing the 12.5-MHz spectrum. Figure 4.3 shows a 12-cell layout with a mobile at the boundary of a cell.

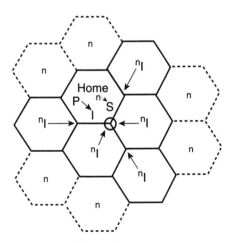

Figure 4.3 Cell Interference Diagram

The C/I can be written as:

$$C/I = kR^{-4}/(k(M-1)R^{-4} + k(2M)R^{-4} + k(3M)(2R)^{-4} + k(6M)(2.6R^{-4})) \qquad 4.19$$

where   M = total number of SS channels   [3]

k = constant

R = radius of cell

-4 = power loss factor

The first term in the denominator is interference from undesired users in the home cell; the second is from interferers in the adjoining cells 2 and 3; the third term is from cells 4, 5 and 6; and the last term is from cells 7 through 12, which are 7.6 R away.

$$C/I \approx 1/(3.312M - 1) \qquad 4.20$$

Given that E/I = 7 dB, B = 1.25 MHz, and R = 8 Kb/sec, solving for C/I in equation 4.19, C/I equals .032(=-15dB). M can be solved in equation 4.20 and substituting all values:

M = 10 channels/cell

Obviously, this performance is not that great. With appropriate power control, interference from all the adjoining cells can be ignored. When k = 1, equation 4.19 reduces to:

$$C/I = R^{-4}/(M-1)R^{-4} \qquad 4.21$$

$$C/I = 1/(M-1)$$

Recalculating,

M = 32 channels/cell

Advantages of CDMA [8]

Although one of the major problems with CDMA is the near-far problem, the following list comprised many advantages of CDMA as a multiple access technique.

1. CDMA is the only multiple access technique that takes advantage of the voice activity cycle to increase capacity.
2. Only a correlator is necessary at the receiver as opposed to an equalizer, which simplifies cost.
3. Only one radio is needed at each cell site, saving money.
4. No guard time is necessary in CDMA.
5. Less fading occurs because of the wideband used.

6. Soft capacity: all users share one channel; additional users can be added with minor degradation. For example: Given a 40-channel system, adding one more user changes the C/I by:

$$10\log(41.40) = .24 \text{ dB}$$

4.3 HYBRID CDMA/TDMA SYSTEM

Both the TDMA and the CDMA components of the proposed hybrid multiple access technique enable voice and data service with different bit rates, allowing a flexible system. The CDMA component offers the advantages of characteristic frequency and interferer diversity and soft capacity. Even though the TDMA component requires overheads, it makes it possible to use joint detection (JD) techniques for CDMA, resulting in increased system capacity compared to conventional detection techniques. Plus, it is possible to have a cluster of one without needing soft handover and the resultant signaling effort between the different base stations. The problem with the technique is that due to time variance and multipath propagation of the mobile radio channel, both the intersymbol interference (ISI) and the multiple access interference (MAI) occur at the receiver.

Two types of receivers can be used with this multiple access scheme: RAKE and joint detection. RAKE treats ISI and MAI as noise distinguishable from signal and also exploits benefits of voice activity monitoring and cell sectorization. However, it requires a tight power control and soft handover. The second type of receiver is the Hybrid scheme.

4.3.1 How Does Joint Detection (JD) Work?

Given: B = bandwidth; K = number of users that are simultaneously active—mobile stations are separated by their different user codes; each mobile station (MS) has one transmit antenna; and transmitted signals receive uplink over K_a receiver antennas. There are therefore, $K*K_a$ radio channels. Each K user transmits in bursts of duration of T_{bu} Figure 4.4 shows the scheme.

The scheme has two data blocks, user-specific midamble for channel estimation, and a guard interval.

Each block has N data symbols of duration T_s and each symbol has Q "chips" (terminology is spread spectrum to differentiate between data and the coding used on the data). $T_c = T_s/Q$ (user specific). The midamble has L_{mid} chips and a guard duration of T_g. The bursts of K users are *simultaneously* active in the *same band* and are synchronized at the receiver except for a timing error, a fraction of symbol interference T_g. By allocating a time slot to each of K tuple of bursts, the TDMA feature is introduced into the JD C/TDMA system and allows for a mixture of voice and data.

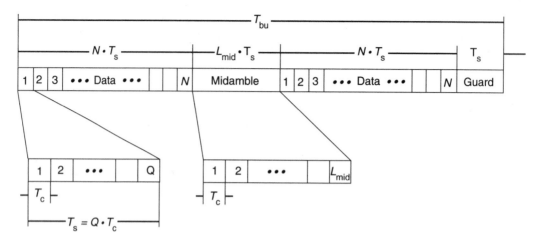

Figure 4.4   Proposed Hybrid Scheme

4.3.2 Single Cell E_b/I_o Performance

The following assumptions are made:

- Coherent receiver antenna diversity is used
- Stationary noise is Gaussian with zero mean
- All variations due to path loss and shadowing are dealt with by perfect power control
- Rayleigh fading
- Two antennas with independent fading process
- 4PSK modulation
- Codes are nonorthogonal
- Rate is 8 kbits/sec
- BER equals 10^{-3}

The parameters are summarized in Table 4.3.

Figure 4.5 shows the average BER probability versus the average E_b/N_o (SNR) per antenna per bit at the receiver for two scenarios: bad urban (BU) and rural area (RA), and two velocities: 30 km/H and 150 km/H. The number of active users varies: 2, 4, 6, 8. With increasing number of users, the E/I required for a specified BER increases. This demonstrates the CDMA characteristic of soft capacity. It appears from the simulation that E/I of JD C/TDMA system is better for BU at 30 km/H than RA at 150 km/H.

Table 4.3 Parameters of the JD-C/TDMA Mobile Radio System for the Simulations

User bandwidth		B	1.6 MHz
Antennas		K_a	2. independent
Burst structure	burst duration	T_{bu}	0.5ms
	midamble	L_{mid}	268
	guard interval	T_g	30μs
	data symbol per data block	N	24
	symbol duration	T_s	7μs
	chips per symbol	Q	14
	chip duration	T_c	0.5μs
	size of data symbol alphabet	M	4 (4PSK)
convolutional encoder:	constraint length	K_c	5
	rate	R_c	1/2
interleaver:	interleaving depth	I_D	4 bursts
	time gap between the beginning of two consecutive bursts of the same user	D_{bu}	6 ms
	resulting net data rate per user	R	8 kbps
filters:	transmitter filter		Butterworth filter of order 4
	receiver filter		Butterworth filter of order 10

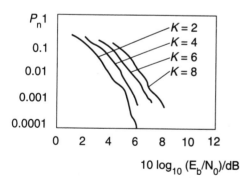

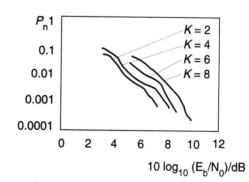

P_b versus E_b/N_0 of the uplink of the JD-C/TDMA system

(a) BU channel model, v=30km/h   (b) RA channel model, v=130km/h

Figure 4.5 Bit Error Probability Versus Average E_b/N_o

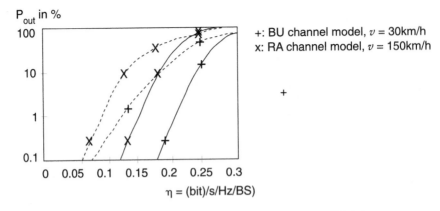

P_{out} versus η of the uplink of the JD-C/TDMA system without FH (...) and with FH(—)

Figure 4.6 Probability of Outage Versus Efficiency

4.4 CELLULAR SPECTRUM EFFICIENCY

Spectrum efficiency, η = bit/s/Hz/BS is determined in the interference limited case for uplink (base station to mobile). BER = 10^{-3} is assumed for good service quality. Also, the system is evenly loaded, or same average number of callers in every cell. Tolerating a certain probability for outage, such as 1%, the cellular spectrum efficiency is given by:

$$\eta = (K_{ave} * D_{bu}) * R/(T_{bu} * B) \tag{4.22}$$

Figure 4.6 shows what the efficiency versus a probability of outage would be under several circumstances: with and without frequency hopping; bad urban environment; rural environment; different velocities. For example, if probability of outage is 1%, frequency hopping applied, if the test is in a rural area, and if the mobile is moving at 30 km/H, then efficiency equals .2 bits/s/Hz/BS. Given the same probability but moving at 150 km/H gives an efficiency of .14 bits/s/Hz/BS. As a comparison, GSM offers an efficiency of .13 bit/s/Hz/BS over a four-cell area. Slow frequency hopping (number of bits per hop as opposed to hops per bit) is employed to improve C/I. This is used in combination with interleaving and coding. Interleaving is used to distribute the effects of burst errors by transmitting bits of a given code block over time by interleaving them among bits of other code blocks.

4.5 CONCLUSION

This chapter analyzed generic capacity, first by defining spectral efficiency, then by deriving the capacity of different systems. Specifically, it was shown that digital systems are inherently more efficient than analog system because of lower susceptibility to cochannel interference. Multiple access techniques, especially CDMA, provide greater capacity than analog FM. CDMA is particularly effective because of the number of users that have access to the entire spectrum at any given time due to the different coding sequences used. As the limitation is due to interference, different techniques, such as antenna sectorization, limit the user interference giving CDMA greater power to accommodate increasing number of users.

In the hybrid multiple access scheme, CDMA/TDMA, it is possible to offer voice and data with different bit rates and still have good efficiency. With increasing number of users, the SNR, or E_b/I_o gradually degrades the system. This defines soft capacity. The cellular spectral efficiency was shown to be superior to that of GSM, which has a cell cluster of four, compared to the hybrid's cluster of one, without soft handover. In the design of future systmes, one must choose the system with the greatest efficiency for a particular application.

Appendix to Chapter 4

Lee [2] defines a parameter called radio capacity. Frequency reuse in cellular radio, although it improves system capacity, is responsible for cochannel interference (C/I) from the cochannel cells. Given interfering cell surrounding the site, six in number, with cell radius R, and cochannel distance D, a minimum D/R ratio must be maintained to avoid cochannel interference:

$$q = D/R \qquad\qquad\qquad A.1$$

The interference comes from the six interferers (cells):

$$C/I = C/ \sum_{k=1}^{6} I_k \qquad\qquad\qquad A.2$$

Let

$$C = \alpha R^{-\gamma}$$

$$I = \alpha D_k^{-\gamma}$$

γ = environmental path loss slope

α = constant

So:

$$C/I = R^{-\gamma}/ \sum_{k=1}^{6} D_k^{-\gamma} \qquad\qquad\qquad A.3$$

The cells are considered uniform so:

$$\sum_{k=1}^{6} D_k^{-\gamma} = 6D_k^{-\gamma}$$

$$C/I = R^{-\gamma}/6D_k^{-\gamma}$$

Substituting:

$$q = D/R = (6(C/I))^{1/\gamma} \hspace{6cm} A.4$$

N equals the number of cells in a cluster, or reuse, pattern. From geometry of hexagons

Radio capacity is defined:

$$m = B_t/B_c*N \hspace{7cm} A.5$$
$$N = q^2/3$$
$$N = (6/3^{\gamma/2}*(C/I))^{2/\gamma}$$

The fourth power path loss is assumed in cellular: = 4.

$$N = \sqrt{2/3(C/I)} \hspace{6cm} A.6$$

Therefore:

$$m = B_t/B_c * \sqrt{2/3(C/I)} \hspace{5cm} A.7$$

Mehrotra [1] defines a parameter modulation efficiency, $\tilde{\eta}_m$

η_m = Total number of channels available/total BW*cluster area

$$\eta_m = B_t/B_c/B_t*N*A \hspace{5cm} A.8$$

where: B_t = total BW

B_c = channel BW

N = number of cells in reuse pattern

A = area of cell

$$\eta_m = 1/B_c*N*A \text{ channels/MHZ/Area} \hspace{3cm} A.9$$

Equation A.9 is called the modulation efficiency because the channel bandwidth is a function of the modulation system. If the efficiency is controlled by geography rather than by modulation, then

$$\eta \alpha 1/Nb_c \hspace{7cm} A.10$$

Equation A.11 shows that efficiency is inversely proportional to the bandwidth, i.e., reduce the bandwidth and/or the area and increase the efficiency. The cluster size N depends directly on the C/I ratio as shown below.

Basically, Lee's derivation is equivalent to Mehrotra's derivation. For example, the U.S. FM cellular system has a total bandwidth of 12.5 MHz and a channel bandwidth of 30 kHz so that

12.5E6/30E3 = 416 channels/cluster

Actually, subtracting 21 channels for signaling leaves 395 voice channels. With a cluster (N) of 7 cells each, there are 56 channels/cell. Given a C/I = 18 dB for this cellular system, then N = 6.5 cells (7 cells).

In Lee:

$$m = B_t/(B_c) \sqrt{2/3(C/I)} \qquad\qquad \text{A.11}$$

$$m = \frac{12.5\text{MHz}}{(30\text{KHz} \sqrt{2/3(10^{1.8})}\,)}$$

$$m = 64 \text{ channels/cell}$$

$$64 \text{ channels/cell} * 6.5 \text{ cells/cluster} = 416 \text{ channels}$$

Subtracting 21 and dividing by 7 equals 57 channels/cell.

To find reuse distance:

From equation A-4,

$$C/I = 1/6(D/R)^4 = 1/6(3N)^2 \qquad\qquad \text{A.12}$$

$$q = D/R = \sqrt{3N}$$

For the FM case, N = 7 so:

$$D = 4.6 \text{ R} \qquad\qquad \text{A.13}$$

where:

D = Reuse distance

= minimum distance between two cells that use the same frequencies

R = cell's radius

REFERENCES

[1] Mehrotra, Asha. *Cellular Radio Performance Engineering.* (location: Artech House, 1994), pp 223-225.

[2] Lee, W. C. Y. "Spectrum Efficiency in Cellular." IEEE Transactions and Vehicular Technology, year:38(2):69-75.

[3] Raith, Krister. "Capacity of Digital Cellular TDMA Systems." IEEE Transactions on Vehicular Technology, year: 40(2):323-331.

[4] Balston, Marcario. *Cellular Radio Systems.* (location: Artech House, 1993), pp 33-36.

[5] Lee, W. C. Y. *Mobile Cellular Telecommunications Systems.* (location: McGraw-Hill Book Co, 1989), pp 47-64, pp 377-384, pp 428-434.

[6] Gilhousen, Klein S. "On the Capacity of a Cellular CDMA System." IEEE Transactions on Vehicular Technology, year: 40(2): 303-311.

[7] Lee, W. C. Y. "Overview of Cellular CDMA." IEEE Transactions on Vehicular Technology, year: 40(2): 290-302.

[8] Lee, W. C. Y. *Mobile Communication Design Fundamentals,* (location: John Wiley and Sons Inc, 1993), pp 296-319.

[9] Blanz, Joseph; Anja, Klein; Markus, Nasshan; Andreas, Steil. "Cellular Spectrum Efficiency of a Joint Detection C/TDMA Mobile Radio System," *Mobile Communications: Advanced Systems and Components*, 1994 International Zurich Seminar on Digital Communications March 1994 Proceedings #783 pp 184-195.

[10] Couch, Leon W. *Digital and Analog Communication Systems,* (location: McGraw-Hill Publishing Co, 1993), pp 338-339, p 690.

[11] Lee, W. C. Y. *Mobile Communications Design Fundamentals*, (location: Howard Sams and Co, 1986), pp 114-120.

Chapter 5

Microcell Design in a PCS Environment

Just as the introduction of Alexander Graham Bell's telephone shaped the way in which society communicates today, the development and integration of personal communication systems (PCS) will shape how society communicates for generations to come. The term PCS has many different meanings, depending upon the context in which it is used, but recently the FCC defined PCS as "a family of mobile or portable radio communications services which could provide services to individuals and businesses and be integrated with a variety of competing networks. The primary focus of PCS will be to meet the communications requirements of people on the move." [1] This definition, while providing some indication to what constitutes PCS, is deliberately inexplicit so as not to restrict design creativity. While an exact definition of PCS may not be known for many years, what is known today is that microcells are the foundation on which PCS will be built.

5.0 MICROCELL DESIGN

Microcell as defined in the PCS environment is different from microcell as defined in the cellular radio environment. One difference between the two microcell definitions is size. In the cellular environment, microcells are generally created as a result of cell splitting.

This cell splitting process is important because it allows for more frequency reuse and therefore additional subscribers can be accommodated. However, for engineering and economic reasons, cell splitting cannot continue indefinitely, and eventually the minimum cell radius, approximately 1 km, is reached. [2] PCS microcells will be much smaller than 1 km and will vary in size, depending upon the type of personal communication system being implemented. In this chapter, PCS microcell design will be approached from two different perspectives: the outdoor, urban or licensed microcell; and the indoor, office or unlicensed microcell. While both types of microcells are discussed, primary emphasis shall be on the licensed microcell.

Before examining the design of PCS microcells, it is important to understand the differences between licensed and unlicensed microcells. Licensed microcells are applicable to those systems that will operate in the 1850–1910 MHz and 1930–1990 MHz frequency bands, using licenses granted by the FCC. From a user perspective, these systems will operate in a manner similar to today's cellular phone system in that users will subscribe to the service and will be charged based upon a monthly subscription fee and a time-of-use fee. Unlicensed microcells are applicable to those systems that will operate in the 1910–1930 MHz frequency band. These systems will not require FCC licensing, but these systems are restricted by Part 15 of FCC regulations which states: 1) these systems will not interfere with any licensed operation, and 2) these systems will not be provided interference protection from any licensed operation. 3) These systems are also restricted in that manufacturers must obtain FCC Part 15 certification prior to marketing the equipment. 4) Since unlicensed PCS equipment will be purchased and owned by the user, monthly subscription fees and time-of-use fees are not applicable. Examples of unlicensed PCS include wireless LANs and office-based cordless telephone systems.

The primary goal of licensed PCS is to provide individuals with voice and data communications from small lightweight communication devices. Because these PCS communicators will be required to operate all day on a single charge and still be smaller and lighter than today's cellular radios, communicator transmitting power must be limited. Transmitting power of tomorrow's PCS communicators is anticipated to be approximately 0.010 watts, whereas today's cellular radios transmit between 0.6 and 4 watts. [2, 5, 6]

5.1 LICENSED PCS MICROCELL DESIGN REQUIREMENTS

As a result of the power requirements cited above, the typical radius of a licensed microcell is expected to be about 100 m and possibly as large as 500 m. [2, 7] Because of this small size, numerous microcells will be required in any PCS environment. If personal communication systems are to be economically feasible, the cost of designing and implementing the microcells and associated base stations must be minimized. One way to reduce microcell site cost is to mount the microcell antenna on existing infrastructure such as a lamp post, a telephone pole or a building. The location and height of the antenna as well as the height, location and shape of surrounding buildings will have a direct impact on signal propagation and is one of the primary design considerations of licensed PCS microcells.

One way to examine radio signal propagation is through the use of computer models. Currently, much research is being directed towards finding suitable computer models, because models that were generated for designing the cellular radio system were based on assumptions that are not valid in a PCS environment. [8] Because PCS antennas may be mounted at heights approximately equal to the heights of surrounding buildings, predicting the diffraction caused by the buildings is a primary concern of many researchers.

In diffraction modeling, the effects of buildings on the signal propagation are often simulated by modeling the buildings as absorbing screens. The diffraction caused by these screens is calculated and used as a basis to estimate cell coverage. Maciel and Bertoni used this type of modeling in addition to the standard line-of-sight (LOS) two-ray model to predict signal propagation in residential and commercial environments. [8] Some of the assumptions used in this computer model included: building height = 8 meters, frequency = 900 MHz and mobile receiver height = 1.8 m. In Maciel's and Bertoni's simulation, transmitter location and transmitter antenna heights of 4 m, 8 m and 14 m were used as variables. Figures 5.1, 5.2 and 5.3 contain some of Maciel's and Bertoni's results when the transmitter antenna height is 8 m. In these figures, the bold lines represent city streets; the dashed lines represent areas above a given signal strength.

"Appropriate to systems used by vehicular traffic, we have computed the signal only at the center of the various streets for a base station that is also in the center of the street. In Figure 5.1 we have indicated with light shading those portions of the street in the vicinity of the base station where the received signal S or S_{LOS} is above the level specified for a base station antenna height H_s = 4m. From the figure it is seen that -80, -90, and -100 dB coverage areas are essentially along the street of the base station, extending to distances of 252 m, 466 m, and 841 m away from the base station, respectively, with only small spillover into the side streets. However, the -110dB coverage area, which extends to a distance of 1057 m along the LOS street, is more complex, having disconnected regions on the next street parallel to that of the base staion."[8]

As can be seen from Figure 5.1 (> -80 dB), the area with the strongest signal strength can best be described as being along the LOS path of the transmitter. It is also obvious that signal propagation perpendicular to the transmitter is extremely limited. With the exception of the > - 80 dB condition, the plots in this figure do not accurately represent the LOS distances. Per Maciel and Bertoni, LOS distances for signal strengths > -90 dB, > -100 dB and > -110 dB are 649 m, 1181 m and 2122 m, respectively.

Figure 5.2 shows the predicted cell shape associated with a signal strength > -110 dB when the transmitter is located at the intersection of two streets. Again, propagation is primarily LOS, but in this case, more area is covered because the transmitter is located at the intersection of two streets. As before, the LOS signal drops below -110 dB 2122 m from the transmitter.

Figure 5.3 shows the predicted cell shape associated with a signal strength > -110 dB when the transmitter is located in the middle of a block that is surrounded by buildings. In this situation, the LOS dominance is removed and the cell shape is more rectangular or diamond shaped than linear.

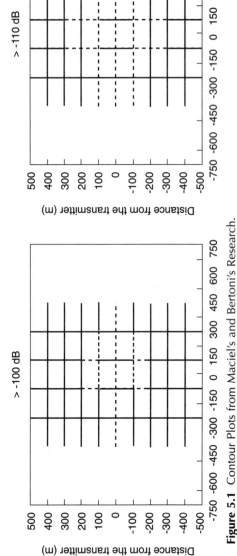

Figure 5.1 Contour Plots from Maciel's and Bertoni's Research. Antenna location is at mid-block and antenna transmitter height is 8 m.

Source: Reproduced from *IEEE Transactions and Vehicular Technology*, May 1994, page 274; or: reprinted from [8].

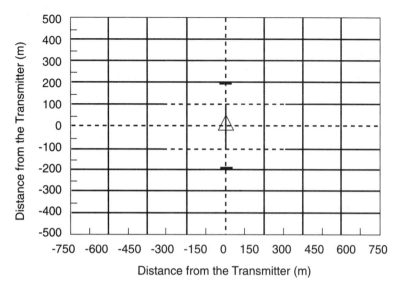

Figure 5.2 -10 dB Contour Plot From Maciel's and Bertoni's Research. Antenna location is at the intersection of two streets and antenna transmitter height is 8 m.

Source: Reproduced from *IEEE Transactions and Vehicular Technology,* May 1994, page 276; or: reprinted from [8]

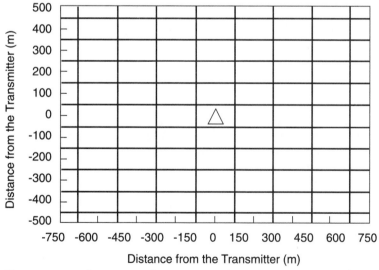

Figure 5.3 -10 dB Contour Plot From Maciel's and Bertoni's Research
Antenna location is in the backyards of a block and antenna transmitter height is 8 m.

Source: Reproduced from *IEEE Transactions and Vehicular Technology,* May 1994, page 276; or: reprinted from [8].

Another model that has been used to predict radio wave propagation is based upon determining the "local mean attenuation" (LMA) for streets in which LOS coverage is applicable and for cross streets in which diffraction coverage is applicable. [2] In this model, LMA represents averaged values of propagation loss that were obtained as a result of sampling the received power at various points along the transmission path. When LMA values were used to predict microcell coverage for a transmitter placed near the intersection of two streets in Manhattan, the results showed due to LOS propagation, that coverage resembled a diamond with elongated axes along the main streets. Although this modeling technique is different from the modeling technique referenced earlier, it is interesting to note that Figures 5.2 and 5.3 also indicate coverage that is comparable to a diamond shape.

This diamond-shaped cell is important because it provides a pattern for laying out microcell coverage. [2] Just as hexagonal-shaped cells are used to represent the coverage of cellular radio systems, it may be possible to use diamond-shaped microcells to plan PCS coverage. Figures 5.4 and 5.5 are intended to demonstrate how diamond-shaped microcells would be used to plan PCS coverage. The thin lines represent the street pattern, and the bold lines represent the microcell shape. The size of the diamond microcells is not based on any objective measurements, but rather is based upon qualitative analysis of the modeling results mentioned above. The microcell shapes in Figures 5.4 and 5.5 presuppose that the transmitting antenna is located near the intersection of two streets but not directly in the center of the intersection. Consequently, the diamond microcell has an elongated axis due to stronger LOS propagation down one street.

Figure 5.4 is an example of how diamond-shaped microcells could be used to represent coverage in the initial stages of PCS implementation. Because PCS must be demonstrated to be economically feasible before coverage becomes widespread, initial coverage may be only along principal avenues where business activity is high and where demand for service is expected. As the system becomes profitable and as demand for PCS increases, coverage would expand.

Figure 5.5 represents coverage in an intermediate stage of PCS implementation. In the intermediate stage, PCS coverage is not complete, but key areas in which demand is anticipated are covered. Although diamond-shaped microcells are a convenient way to plan coverage, it is important to remember that the diamond shape is based upon models that used a rectangular street grid. As a result, using a diamond to approximate microcell coverage should be limited to geographic areas in which the street pattern is rectangular.

Another aspect that must be considered when designing a microcell is the location of the microcell transmitter and receiver. In cellular radio, the transmitter and receiver are typically located at the base of the antenna tower. Because PCS antennas may be mounted on lamp posts or telephone poles, locating the transmitter and receiver in the proximity of the antenna may not always be a feasible option. An alternative is to locate the transmitter and receiver at a remotely located base station and connect the base station to the antenna via optical fiber. In this configuration, the transmitting system at the antenna site would simply consist of an optical receiver, a bandpass filter and a power amplifier. The receiving system at the antenna site would consist of a bandpass filter, an amplifier and an optical source that would be modulated by the received signal. [2] One advantage of this con-

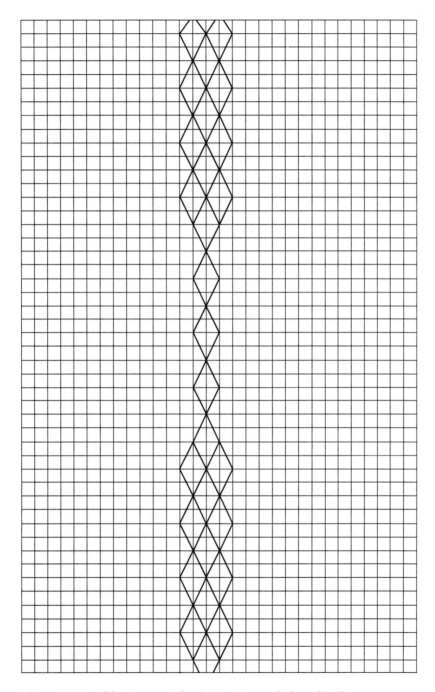

Figure 5.4 Possible Coverage Plan Using Diamond-Shaped Cell
(Initial stage of PCS implementation)

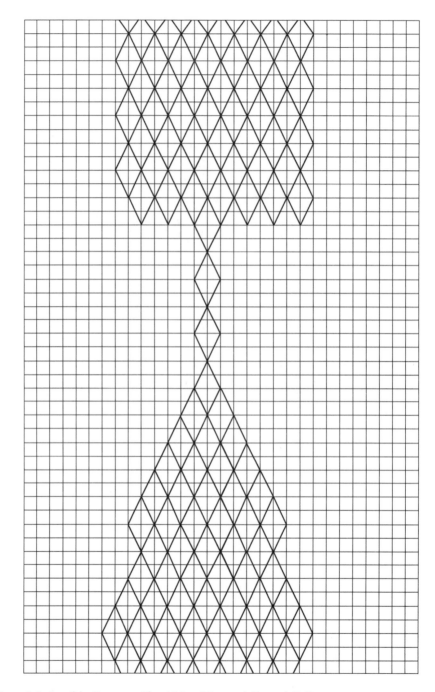

Figure 5.5 Possible Coverage Plan Using Diamond-Shaped Cells
(Intermediate stage of PCS implementations)

figuration is that the signal processing required to extract the baseband information from the bandpass signal could be performed at the base station, thus minimizing the space required for equipment at the antenna site. Studies using this type of configuration have been performed for cellular systems operating between 800–900 MHz. The results indicate that for a base station located 12 km from the antenna, up to sixteen 30-KHz channels can be handled by one antenna site if a 1 W RF amplifier and a 1 mW laser are used. [2] The above findings are significant because they indicate that the signal processing for many microcells could be handled at one centralized location. This is important because it reduces the number of suitable remote locations that must be located and built or leased. As a result, base station investment costs and operating and maintenance costs are reduced. Systems in which the signal processing equipment for many microcells is centrally located are said to have a centralized architecture. [9]

While centralized architecture allows for reducing the amount of equipment at the antenna site, it has a disadvantage in that optical fiber must be installed between the antenna sites and the base station. In some situations, it may be cost prohibitive to install optical fiber between these locations. In these instances, the transmitter, receiver, and signal processing equipment will be co-located with the antenna. Systems in which signal processing equipment is co-located with the antenna are said to have a distributed architecture. [9]

The above paragraphs provided background information about PCS and addressed two of the many factors that must be considered in designing a licensed PCS microcell, cellular coverage and network architecture. Because of the low antenna heights associated with PCS, microcell coverage is largely a function of existing infrastructure. Predicting microcell coverage with computer models is a difficult and inexact science, because buildings vary in shape and size and because it is difficult to include these type of variables in a computer model. Nevertheless, computer modeling is an important and powerful tool and should be used in conjunction with actual measurements. The decision whether PCS should have a centralized or distributed architecture will most likely be based upon economic and engineering considerations. In all probability, tomorrow's licensed PCS will contain both types of architectures.

5.2 UNLICENSED PCS MICROCELL DESIGN REQUIREMENTS

Most, if not all, unlicensed personal communication systems will be located inside buildings. Therefore, the unlicensed microcells will have signal propagation characteristics that are different from those observed in cellular radio or licensed PCS environments. In addition to the propagation differences, microcell size and user requirements for unlicensed PCS will be varied and unique. Propagation characteristics, cell size and user requirements are just a few design considerations that must be taken into account when designing an unlicensed PCS microcell.

Successful integration of unlicensed PCS equipment into a building requires that indoor radio propagation characteristics such as multipath fading, delay spread, and the effects of antenna placement and antenna configuration be clearly understood. Research by AT&T [2] showed that multipath fading does occur in buildings, but the corresponding average delay spread for these signals can be measured in nanoseconds rather than microseconds. This is important because the small delay spread translates into greater possible transmission rates—an essential requirement if high-speed wireless LANs are to become a reality. The AT&T research also showed that the shape of the building is very important in determining proper antenna configuration and placement. For example, in areas that are much longer than they are wide, such as long halls with offices on each side, a distributed antenna system reduced the delay spread and lessened the power loss when compared to a system with one centrally located antenna. In the above test, the distributed antenna system consisted of a coaxial cable with monopole antenna elements located every 6 m along the hallway ceiling of a 115-meter building. Although this type of arrangement increased the number of multipath signals, it also reduced the delay spread, and that is why this configuration is beneficial. [2]

Because unlicensed PCS will be installed within the confines of buildings, unlicensed PCS microcells will be much smaller than licensed microcells. In an effort to differentiate between licensed and unlicensed microcells, unlicensed microcells are sometimes referred to as picocells, nanocells and femtocells. There are no consistent definitions of the size of a picocell, nanocell or femtocell other than what can be inferred from the prefixes of the terms, but generally speaking, picocells have radii between 10 and 30 m, nanocells have radii around 5 m, and femtocells have radii of 2 or 3 meters. [2, 10] The obvious advantage of small cells is that they allow for more frequency reuse, and therefore a large number of users can be accommodated with limited frequency spectrum.

Finally, the function of *unlicensed* PCS will be different than that of licensed PCS or cellular systems, and this must be taken into account when designing unlicensed microcells. As stated earlier, the types of systems anticipated for unlicensed PCS include wireless LANs, wireless PBXs and office-based cordless telephone systems. In these systems, the users are traveling relatively slowly or, in some cases, are stationary. Unlicensed microcell design can take advantage of these user characteristics and focus on designing a system tailored to meet the user's specific needs. For example, if a wireless LAN is to be installed in a specific building, and if it is known that all LAN users will be in offices and that LAN service is not required in the hallway, then a system design can be based upon providing coverage only to the office areas. Providing coverage to only the required areas is important because it conserves monetary as well as frequency spectrum resources. Because unlicensed PCS will be purchased by the user and designed around the user's requirements, some experts feel that unlicensed PCS may be the most profitable area of PCS development. [1]

5.3 CONCLUSIONS

As is evident from the discussion above, designing a microcell in a PCS environment is very challenging because so many variables influence the design. Some of the variables that must be considered include signal propagation, antenna heights, building heights, transmitter configurations, receiver configurations, and user requirements and expectations. Equally apparent from the previous paragraphs is that microcells will be the backbone or foundation on which personal communications systems are built, and, therefore, proper microcell design is imperative. Although many years will pass before the PCS envisioned in today's laboratories become a reality in tomorrow's world, PCS's eventual impact on society's day-to-day activities will be nothing less than monumental.

REFERENCES

[1]   Kobb, B. Z. "Personal Wireless." *IEEE Spectrum*, June 1993: 30(6):20–25.

[2]   Greenstein, L. J. et al. "Microcells in Personal Communications Systems." *IEEE Communications Magazine,* December 1992: 30(12):76–87.

[3]   Osbrink, N. K. and J. F. Girand. "New Offerings Put Pressure on Public-Use Spectrum," *Microwaves & RF*, April 1991: 30(4)45.

[4]   Colmenares, N. J. "The FCC on Personal Wireless." *IEEE Spectrum*, May 1994: 31(5)46.

[5]   Cox, D. C. "Wireless Network Access for Personal Communications." *IEEE Communications Magazine*, December 1992: 30(12):96

[6]   Lee, W. C. Y. *Mobile Cellular Telecommunications Systems*. (New York: McGraw-Hill, 1989)

[7]   Xia, H. H., et al. "Radio Propagation Characteristics for Line-of-Sight Microcellular and Personal Communications." *IEEE Transactions on Antennas and Propagation*, October 1993: 41(10)1439–1447.

[8]   Maciel, L. R., and H. L. Bertoni. "Cell Shape for Microcellular Systems in Residential and Commercial Environments," *IEEE Transactions on Vehicular Technology*, May 1994: 43(2):270–278.

[9]   Sarnecki, J., et al. "Microcell Design Principles." *IEEE Communications Magazine*, April 1993: 31(4)76–82

[10] Wayner, P. "Stretching the Ether." *BYTE*, February 1993: 18(2):159(4)

Chapter 6

The Hand-Off Problem in Cellular Radio Systems

In cellular PCS systems, the number of calls arriving in a base station or switch coverage area is random and time varying. As users with mobile telephones move between cells, large variations in the number of telephones in a base station coverage area occur in a matter of minutes. Therefore, one can see that one of the biggest problems in cellular systems is the mobility of users. Variations in vehicular traffic, build-up mobile telephones in a cell, their speed and their direction of movement—these relate directly to the call arrival and hand-off rate. In this chapter, we investigate the hand-off problem. Hand-off problems arise in cellular mobile systems when a user moves from one cell site to another and is of particular interest in microcell design. The hand-off problem is of particular concern in Low Earth Orbit (LEO) satellite systems, since a subscriber may pass to many satellites for the duration of a call.

First, we discuss the hand-off problem and then describe the hand-off process of various cellular systems already in place or being proposed. One of the best ways for an engineer to attack this problem is by modeling. Since hand-off depends on many factors, such as size of the cells, boundary length, signal strength, man-made noise, fading and other forms of interference and since most of these factors are due to the mobility of the users, we intend to model hand-off in two ways. One model will be based on the shape and size of the cells, and the other based on statistical theory. These two subjects will be addressed as a model based on geometry and statistical modeling.

6.0 THE HAND-OFF PROBLEM

After the setup of a call, the continuation of the conversation depends on several factors. Such factors may include the following: reliable detector of conditions that indicate the need of a mobile for a hand-off; identification of alternative cell sites for possible service; timely exchange of supervisory signals; and deployment and availability of communication resources. When a mobile is engaged in a call, it will frequently move out of the coverage area of the base station with which it is in communication, and unless the call is passed to another cell, it will be lost. Effective and reliable handover is not only highly desirable from the user's point of view, but also essential in the control of cochannel interference and therefore the maintenance of the cell plan, especially as the cell size is reduced. As one can see, hand-off will result in forced termination of calls which in turn will lead to interruption in service. What do we mean by hand-off of a call? Because of the mobility of a user from cell to cell, it is required that a call that already has been established should not be interrupted. In other words, the path (channel) of the call should be changed while the call is in progress and have little or no noticeable effect to the user. The user may or may not participate in the process of hand-off. There are some kind of measurements that will define the right time for hand-off. Some of these factors include:

- Signal strength
- Signal phase
- Combination of the two
- BER
- Distance

Hand-off is needed generally in two situations: when the mobile is at the boundary of one cell entering another cell, in which case one of the factors above will be met (depending on the system), for example, if the receiving signal is weak, which may happen if the mobile is moving from one cell to another; or when the mobile is reaching a signal strength gap within the cell. Figure 6.1 illustrates this situation.

Until now, the most popular way of determining the need for hand-off is the signal strength. Throughout this chapter, we will use the signal strength as the factor that will determine the need for hand-off. In general, there are two ways to determine the need for hand-off based on the received signal. The first is based on the signal strength, and the other is based on the carrier-to-interference ratios.

Two systems are in use worldwide at this moment, the analog cellular system and the digital cellular system. Each system uses a different hand-off procedure.

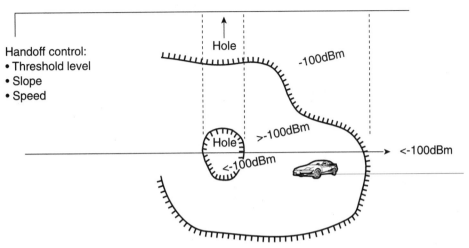

Figure 6.1 Areas That Will Initiate Hand-off

6.1 HAND-OFF IN DIGITAL CELLULAR SYSTEMS

In most of the digital cellular systems, hand-off is performed with the assistance of the mobile. Two most popular hand-off procedures are the one used by Europeans and the one used by Americans.

In the European system, called Global System Mobile (GSM), three parties are involved in the call setup, continuation hand-off procedure: the mobile, the Base Station Controller (BSC) and the Mobile Switching Center (MSC). During a conversation, if the mobile moves from one cell to another, its measurements will indicate that the signal is becoming weak in the present cell site to which it is linked, but the signal strength from another cell within the same BSC is satisfactory. At the same time, BSC will indicate similar result.

If both user's and BSC measurements indicate that the signal strength of some other neighboring cell is better, then BSC decides to transfer the call to the other cell channel, informing the mobile to tune to the new channel assigned. This operation assumes that the two cells are both controlled by the same BSC. If that is not the case and the other cell is controlled by another BSC but still within the same MSC, the process is the same. The only difference is that again the channel will change. The last case is when the user comes to a cell that is controlled by different MSC and BSC. In this case, the MSC through which the call was initiated will extend the call link through the new MSC and a convenient MSC but will retain supervision of the call. Figure 6.2 shows a schematic procedure for handover.

In the American system, when the cell site senses that the signal strength is below a certain level, it alerts the Mobile Telephone Exchange (MTX). The MTX commands adjacent cells to make measurement on the channel in use by the mobile. In turn, the cells sites

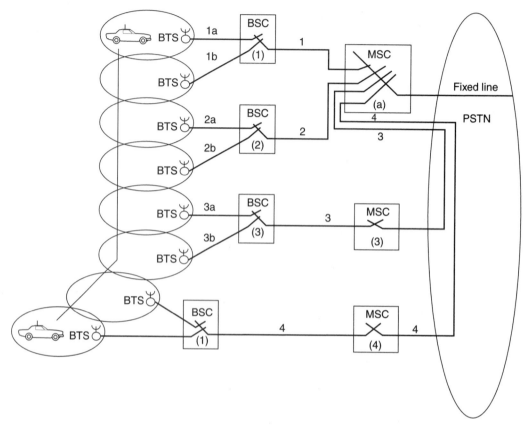

Figure 6.2 Hand-off of a Roaming Mobile for GSM System

request the mobiles that have been served at the time to make an estimate of the signal strength on a specified channel. All this information is gathered at the MTX; as a result, MTX knows which is the best cell site for the mobile. MTX informs the cell site that the mobile is using to send a hand-off message and also alerts the new cell site. The old cell site transmits the message to the mobile; the message includes the new channel assignment, power level, etc.

After receiving the hand-off message, the mobile turns on the signaling tone for 50 ms, turns off the signaling tone and the transmitter, adjusts the power level, retunes to the new channel and sets the transmitter and receiver to the digital mode. Once the mobile is synchronized, the transmitter is turned on. The two processes described above are used with Frequency Division Multiple Access (FDMA) and Time Division Multiple Access (TDMA) techniques.

A third technique, still in the experimental state, uses a new way to treat hand-off process. This technique is called Code Division Multiple access (CDMA), and the hand-off process that is used is called soft/softer hand-off. In this case, the mobile constantly monitors the signal strength from nearby stations. The mobile selects the best communications path to maintain an optimum signal quality. As the mobile approaches a cell boundary, a link is established with another based station in a neighboring cell, while the original connection with the old cell is maintained. This ensures a smooth transition between cells and requires no action on the part of the mobile. The cell is now carried by two, or possibly more, base stations. The new cell site switches to the code the mobile is using. When the user establishes a firm connection with the new cell and is satisfied with the results, the old cell is dropped.

6.2 HAND-OFF IN ANALOG CELLULAR SYSTEMS

The hand-off process in analog cellular systems starts when the mobile signal level, received at the old site, falls below a preassigned threshold. Once this condition is met, the cell serving the mobile notifies the MTSO to check among the adjacent cells where the signal of the mobile is received best. When the best new cell is identified, the serving cell site is notified of the new channel number to which the mobile should tune for hand-off. As the mobile resumes to the new channel, hand-off has essentially taken place. The procedure for hand-off is described below.

The Mobile Telephone Switching Office (MTSO) has the location information of the mobile all the time. When the carrier drops below a certain level, the MTSO decides to switch over the present call from the old cell site to the new cell site. The MTSO sets an idle noise channel at the receiving cell site and informs the new cell site to switch on its transmitter. The message is sent to the mobile through the active serving cell site informing it of its new voice channel designation. The mobile turns off the supervisory tone from the old voice channel, which is interpreted at the MTSO as going on-hook. The mobile also returns to the new channel and transports the supervisory audio tone (SAT) found there. This is recognized by the MTSO as a successful completion of the hand-off sequence. The MTSO reconfigures its switching network and connects the landline party to the mobile through the new voice channel and landline trunk. The entire hand-off process takes about 0.2 seconds in the form of brake-before-connection. Hand-off process is shown in Figure 6.3.

Becuase hand-off will be needed, some channels must be reserved for this purpose. If we assume that there are C_r channels in a cell reserved for hand-off, and the total capacity of the cell is C, then the available channel for use by the mobiles of the particular cell is:

$$Cav = C - C_r \qquad\qquad 6.1$$

From Equation 6.1, one can say two things: New calls will be blocked if the number of channels in use is greater than Cav; hand-off attempts will be blocked if the number of channels used is C.

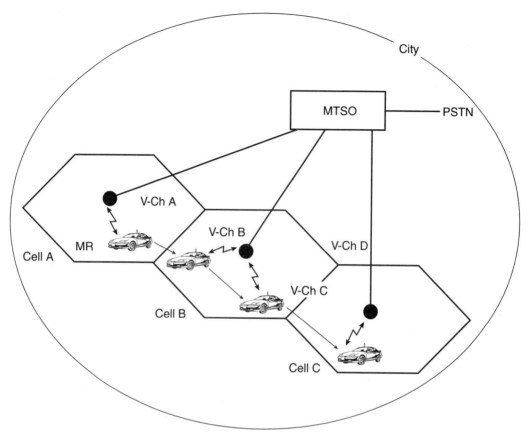

Figure 6.3 Hand-off Process in Analog Systems

6.3 ANALYSIS OF HAND-OFF

Cellular systems in the future should evolve into smaller cells (microcells) in order to cope with the daily increase in number of users. In this analysis, we will use this fact with the assumption that cell coverage areas are designed to carry equal amounts of traffic and cell sizes are equal. Based on these assumptions, we believe that simple geometry is one way of predicting hand-off rates. Additional assumptions are that mobiles are distributed uniformly within the cell and that the direction of travel of mobiles is uniformly distributed over $[0, 2\pi]$.

Under these assumptions, the hand-off rate should be proportional to the total boundary length. It is obvious that the hand-off rate increases as the cell size decreases.

6.3.1 Geometric Modeling

Today's cellular systems often have smaller cells in order to cope with increasing demand. One can use simple geometry to predict hand-off rates in the smaller cells. Since coverage areas are often designed to carry equal amounts of traffic, equal-sized cells, as shown in Figure 6.4, carrying homogenous traffic is a useful assumption.

If the number of channels per cell is N and each cell is a square with side a, then the total area of a cell is a ^{2}.

The channel per unit area then is

$$\rho' = \frac{N}{a^2}$$
6.2a

and the boundary length per unit area is

$$\frac{4a}{a^2} = \frac{4}{a}$$
6.2b

As the number of users in the cell increases, we must increase the number of channels per cell (or with the same talking channel per unit area) to be able to meet the new demand. New technology and spectrum allocation increase the number of channels per cell by factor of G, so the total number of channels in a cell becomes GN. Then, the channel per unit area becomes:

$$\rho_{\text{tmpr}} = \frac{GN}{a^2}$$
6.3

but still not enough to cope with the required demand.

One way to meet the new demand, especially in PCS sytems, is to reduce the size of the cell (perform cell splitting). If the original square side is a, then we reduce the side by some factor F to. $\frac{a}{F}$.

$$\rho_{\text{new}} = \frac{GN}{(a/F)^2}$$
6.4

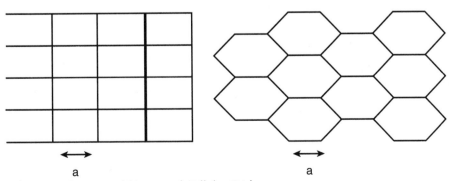

Figure 6.4 Square and Hexagonal Cellular Grids

As a result, the number of channels per unit area increases by a factor of GF^2. For a hexagonal grid, we have

$$\text{Area} = \frac{1}{4}na^2\cot(\frac{\pi}{\eta}) = \frac{1}{4}6a^2\cot(\frac{\pi}{6}) = \frac{6}{4}a^2(1.732) = 1.5\sqrt{3a}^2 \qquad 6.5$$

$$\rho_n = \frac{N}{1.5\sqrt{3a^2}} \qquad 6.6$$

$$\rho_{n,\text{new}} = \frac{GF^2N}{1.5\sqrt{3a^2}} \qquad 6.7$$

As mentioned earlier, by splitting the cells in order to accommodate the new demand, we also increase the hand-off rate.

Again, for the square shape of the cell, the total cell boundary length per unit area increases from $\frac{4a}{a^2}$ to $\frac{4aF}{a^2}$ by a factor F. With the increase in call density discussed earlier, we have an increase in hand-off density by a factor of GF^3. The number of calls per unit area increases as GF^2, and the number of hand-offs per unit area increases GF^3; therefore, the hand-off rate per call increases as F.

Figure 6.5 illustrates the effects of cell splitting by a factor F.

The above development assumes that there is no change in user mobility or increase in demand. If the existing demand is dominated by vehicular mobiles and the increased demand comes from pedestrians, then the above development would not apply.

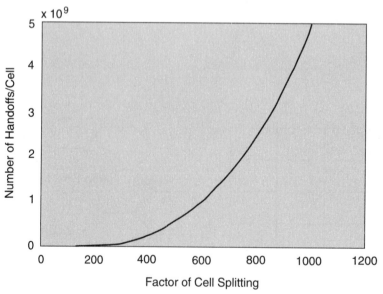

Figure 6.5 Effect of Cell Splitting on Hand-off

The shape of the cell is the next important factor that will affect the hand-off rate. If we assume that the traffic is homogenous, the hand-off rate will depend only on the length of the boundary. Based on this, the best shape we could use is the circle, which will have $\frac{2}{\alpha}$ boundary length per unit area(α is the radius of circle), a factor of 2 smaller than the square shape. On the other hand, hexagonal shape has a factor of $\sqrt{3}$ smaller than the square shape. Now, if we consider that the traffic is directional, then the hand-off rate will depend not only on the length of the boundary, but also on its orientation, as shown in Figure 6.6

Assume the following: The number of vehicles crossing a unit length of boundary, oriented N-S, is y; the number of vehicles crossing an E-W oriented boundary is x; and also consider rectangular cells of area A. If we choose a rectangle that is oriented at an angle (from E-W direction), as shown in Figure 6.6, with sides a, b, then ab = A. Then, if we define H to be the hand-off rate and using simple trigonometry, the hand-off rate is:

$$H = a(x\cos\Theta + y\sin\Theta) + b(x\sin\Theta + y\cos\Theta) \qquad 6.8$$

With the cell area fixed at A, the number of calls in the cell is fixed. Then for a fixed Θ, to minimize the hand-off rate H, we should choose:

$$a = \sqrt{A}\sqrt{\frac{x\sin\Theta + y\cos\Theta}{x\cos\Theta + y\sin\Theta}} \qquad 6.9$$

$$b = \sqrt{A}\sqrt{\frac{x\cos\Theta + y\sin\Theta}{x\sin\Theta + y\cos\Theta}} \qquad 6.10$$

$$h = 2\sqrt{A}\sqrt{\frac{xy + (x^2 + y^2)}{\sin\Theta\cos\Theta}} \qquad 6.11$$

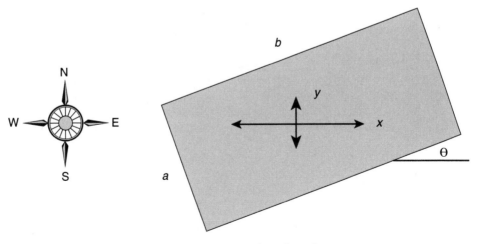

Figure 6.6 Rectangular Cell Orientation at an Angle Θ from the Aaxes

which is minimized for Θ = 0, if the rectangle is oriented in the E-W direction, Figure 6.7 depicts the relation of the hand-off rate to the traffic flow through the boundaries. If

$$\Theta = 0$$

$$a^2 = \frac{Ay}{x} \qquad H = 2\sqrt{Axy} \qquad\qquad 6.12$$

then the boundaries are normal to the flow and $b^2 = \frac{Ax}{y}$

$$\frac{b}{a} = \frac{x}{y} \qquad\qquad 6.13$$

This suggests that we must put the longer boundaries in the direction that has a smaller boundary crossing rate.

6.4 HAND-OFF RATE FROM CALL HOLDING TIME AND CELL SOJOURN TIME

In general, the call holding time is exponentially distributed: let γ equal t seconds with mean $\frac{1}{\mu}$. As a mobile traverses a cell, as shown in Figure 6.8, assume that its sojourn

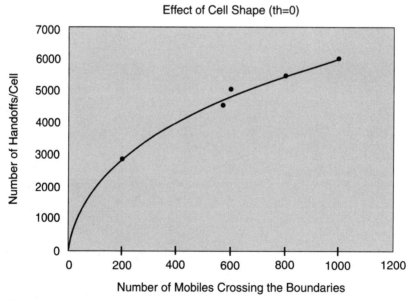

Figure 6.7 Effect of Cell Shape (th = Θ)

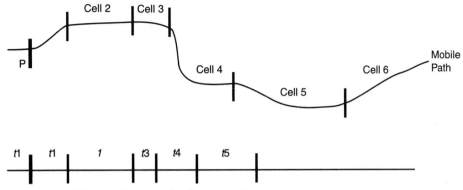

Figure 6.8 Mobile Travel Path and Cell Sojourn Times

times t_1, t_2, ..., are independent, identically distributed random variables with the same mean $\frac{1}{\gamma}$. Then, the average hand-off rate is $\frac{\gamma}{\mu}$.

Consider a mobile traveling on a path, as shown in Figure 6.8, with consecutive cell sojourn times t_1, t_2, ... Assume that the mobile initiates a call at instant P so that the remaining sojourn time in the cell is t'; let H be the number of hand-offs per call. We are trying to find the expected number of hand-offs per call E[H]. Since we have already defined the mobile's sojourn times in the cells, the number of hand-offs that the call will go through depends on the holding time, which is exponentially distributed with mean 1/m. Therefore,

$$\text{E[H]} \ t', t_2, t_3,...] = 1. \int_{t_1}^{t'+t_2} \mu e^{-\mu x} dx + 2. \int_{t'+t_2}^{t'+t_2+t_3} \mu e^{-\mu x} dx + 3. \int_{t'+t_2+t_3}^{t'+t_3+t_4} \mu e^{-\mu x} dx + ...$$

$$= \int_{t'}^{\infty} \mu e^{-\mu x} dx + \int_{t'+t_2}^{\infty} \mu e^{-\mu x} dx + \int_{t'+t_2+t_3}^{\infty} \mu e^{-\mu x} dx + ... \qquad 6.14$$

$$= e^{-\mu t'} \ [1 + e^{-\mu t2'} + e^{-\mu t3'} \ [1 +...$$

If t_1 is identically distributed with probability density function g(t) and the function G*(s) is used as the Laplace transform of g(t), then

$$= e^{-\mu t'} \ [1 + G*(\mu[1 + G*(\mu[1 +... \qquad 6.15$$

$$= \frac{e^{-\mu t'}}{1 - G*(\mu)}$$

To calculate E[H], we need to solve the distribution of t' —the remaining sojourn time in the first cell. If t_1 is exponentially distributed, then the problem will be easier to solve: t' has the same distribution because of the special characteristic of the exponential distribution of being memoryless. In general, if t' has the probability density function r(t), with Laplace transform R*(s), then, using 6.15,

$$E[H|t'] = \frac{R*(\mu)}{1 - G*(\mu)} \qquad\qquad 6.16$$

Using the residual service time,

$$R*(s) = \frac{1 - G*(s)}{s/\gamma} \qquad\qquad 6.17$$

Substituting 6.15 in 6.16,

$$E[H] = \frac{\gamma}{\mu} \qquad\qquad 6.18$$

Therefore, the average hand-off rate is equal to the ratio of the mean call holding time to the mean cell sojourn time. [1]

6.5 STATISTICAL MODELING

The decision to initiate a hand-off process, as mentioned earlier, can be based on the received signal level from the communicating and neighboring base stations, the distance from base stations and the bit error rate. Signal level is one of the most commonly used criteria for an analog system. This level is said to be -100 dBm, for FM 18 dB of carrier-to-noise ratio. Since we decided that the threshold level will be -100 dBm, we must choose a level higher than -100 dBm to initiate hand-off because, of the time that the hand-off will take to be completed and accounting for the mobile speed not to pass the -100 dBm level. In doing so, we generate a problem. Let's assume that we set a small Δp or we do not choose any higher level to initiate hand-off, then there is not enough time to complete the hand-off process and the call will be dropped. Therefore, the parameter Δp should be varied according to the path-loss slope of the received signal strength and the level crossing rate of the signal strength, as shown in Figure 6.9.

Once we decide on Δp, we can calculate the velocity V of the mobile unit, based on the predicted level crossing rate at a - Δp level with respect to the root mean square level, which is at -100 dBm + Δp.

$$V + \frac{\eta\lambda}{\sqrt{2}x(.27)} \text{ ft/s} = \eta\lambda \text{ mi/h} \qquad\qquad 6.19$$

where η is the level of crossing rate (crossings/s) counting positive slopes, and λ is the wavelength in feet.

The level of the received signal will depend on some parameters such as distance from the base station, fading and shadowing.

The instantaneous radio signal at the mobile can be expressed as

$$s(t) = r(t) \, e^{j\psi(t)} \qquad\qquad 6.20$$

and the received signal envelope r(t) can be described as the product of a slow varying signal m(t) and the rapid variation factor $r_0(t)$.

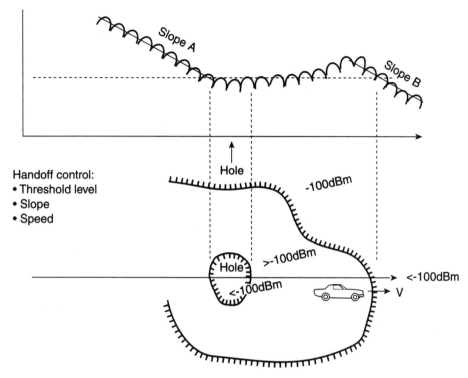

Figure 6.9 Hand-off Algorithms for Necessary and Unnecessary Hand-offs
Source: Reprinted from [2]

$$\Rightarrow r(t) = m(t).r_0(t)$$

Furthermore, we can write an equation as a function of position variable y

$$\Rightarrow r(y) = m(y).r_0(y)$$

$$\Rightarrow r_0(y) = \frac{r(y)}{m(y)}$$

and in the dB notation:

$$r_{0db}(y) = r_{db}(y) - m_{db}(y) \qquad\qquad 6.21$$

Unfortunately, direct analysis of the above variables will not be feasible. So, we will try to study separately the rapid variations variables of $r_0(y)$ by using Rayleigh or Rice probability laws and the slow variation with the log-normal and the normal distribution for $m(y)$.

6.5.1 Rapid Variation On A Linear Scale

When the density of a certain area is such that the mobile receives only multiple reflected and refracted signals without any direct or main component, in general such cases signal a Rayleigh probability density.

$$P(r_0 \leq n_{ro} - k\sigma_{ro}) = 1 - e^{(-(nrok\sigma ro)2/2\alpha2)}$$ 6.22

with mean value and standard deviation specified as

$$n_{ro} = a \sqrt{(\pi/2)}$$ 6.23a

$$\sigma_{ro} = a \sqrt{(2 - \pi/2)}$$ 6.23b

Since the mean for rapid variations has a value of unity; $\alpha = .7979$ and the standard deviation is .5227.

If we take into account that the mobile will also receive a direct component, we can use the Rice probability density:

$$f_{r_o}(r_0) = \frac{r_0}{\alpha^2} e^{(-\frac{r_o^2 + b^2}{2a^2})} I_0(\frac{r_0 b}{\alpha^2})$$ 6.24

(Jean [3], 39)

where $I_0(x)$ is a Bessel function of order zero.

The mean of the probability density function is given by

$$n_{r_o} = e^{-\gamma} \sqrt{\frac{\pi}{2}} a[(1 + 2j) + 2jI_1(j)]$$ 6.25

(Papoulis [4], 75)

where $I_1(j)$ is a Bessel function of order 1 and $j = \frac{b^2}{4a^2}$.

If we assume that the mean is unity (usually the case), then we can assume values for b and solve for α and for the case of unit mean then

$$\sigma_{r_o} = \sqrt{b^2 + 2a^2 + 1}$$ 6.26

b is the direct component, which has the effect of decreasing the standard deviation of the signal envelope as it increases relative to the multipath. For example, for $b = 0$ ($\alpha = .7979$), the standard deviation is .5227 and we have the Rayleigh mode. For $b = .95$ ($\alpha = .30343$) and the standard deviation is .2943. [5]

6.5.2 Slow Variation On A Linear Scale

The slow variation in the receive signal can be described as a log-normal probability distribution linear scale or as a normal probability density when the received signal is expressed on a dB scale. The expression for the probability density and for the distribution is:

$$f_{mdb}(mdb) = \frac{1}{\sigma_{mdb}\sqrt{2x}} e^{(-(\frac{mdb-n_{mdb}}{2\sigma^2_{mdb}})^2)}$$ 6.27

From the above analysis a possible model for the cellular channel can be derived, as shown in Figure 6.10.

6.5.3 Level Crossing Rates And Mean Fading Lengths

Level crossing rates (LCR) are defined as the rate at which the signal at the mobile crosses a certain threshold level in one direction, either with the slope positive or with a negative slope.

The general expression giving the level crossing rate (for positive slopes) for level ψ and stationary time signal $\psi(t)$ is

$$LCR(\varphi = \psi) = \int_0^\infty \psi^f{}_{\varphi\psi}(\psi\varphi)d\varphi$$ 6.28

where $f_{\psi\psi}(\psi.,\varphi.)$ is the joint probability of the average number of level crossing per meter through level ψ.

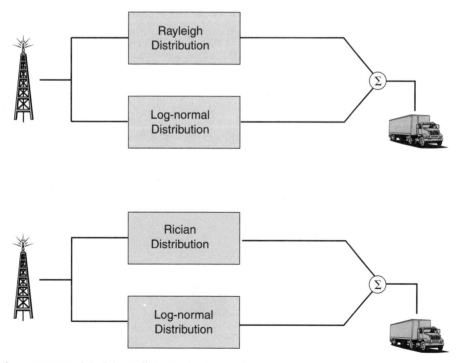

Figure 6.10 Model of the Cellular Radio Channel

In order to find the LCR for the Rice probability law, the average number of level crossings per meter of some threshold R_0 is given by

$$\overline{LCR}(r_0 = R_0) = \frac{f\sqrt{\pi}}{c} \frac{R_0}{\alpha} e^{-\frac{R_0^2+b^2}{2\alpha^2}} I_\theta\left(\frac{R_0 b}{\alpha^2}\right) \qquad 6.29$$

where f is the operating frequency in Hz, c is in m/s (3×10^8 m/s), and the values of b and a are determined by the direct-to-multiple ratio, described as follows:

$$\text{DMR}_{dB} = 20 \log_{10}\left(\frac{b}{\alpha \frac{\sqrt{\pi}}{2}}\right) \qquad 6.30$$

For a level expressed in relation to the signal mean, we can replace R_0 with $\eta_{ro} - k\delta_{ro}$

$$R_0 = \eta_{ro} - k\delta_{ro}$$

and the mean fading length (MFL) under a given signal level is given by

$$\text{MFL}(r_0 = R_0) = \frac{P(r_0 \leq R_0)}{LCR(r_0 = R_0)} \qquad 6.31$$

$$\frac{\int_0^{\eta_{ro}} f_{r0}(r_0) dr_0}{LCR(r_o = R_0)} = \frac{\int_0^{u_{ro}-k\delta_{ro}} \frac{r_0}{\alpha^2} e^{-(r_0^2+b^2)} I_0(r_0 \frac{b}{a^2}) \, dr}{f \dfrac{\sqrt{\pi}R_0 e^{-(R_0^2+b^2)/(2a^2)} I_0(\frac{R_0 b}{a^2})}{C}} \qquad 6.32$$

The next step is to define the appropriate number of crossings in one direction; once this number is reached, the hand-off process will be initiated. This number should be large enough to avoid ping-ponging from one cell to another but not so large as to cause a signal outage and call dropout. [6]

The Model: Assume that a mobile is moving from Station A to Station B, as shown in Figure 6.11.

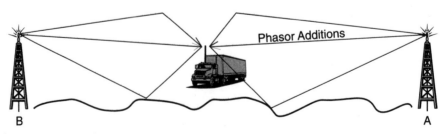

Figure 6.11 Mobile Moving From Station A to Station B

At the vehicle's receiver two signals are arriving, one from station A and one from station B. We can model the signal arriving from station A as a function of distance:

$$A(d) = K1 - K2(\log(D-d)) + U(d)$$ 6.33

and the signal arriving from station B as

$$B(d) = K1 - K2(\log(D-d)) + V(d)$$ 6.34

where D represents the distance between the cells, and the parameters K1, K2 account for the path loss. The shadowing and fading process are included in V(d) and U(d). So, the total signal arriving at the mobile is the difference of the two signals:

$$r(d) = A(d) - B(d)$$ 6.35

Assume that the resultant signal r(d) has the form shown in Figure 6.12.

Assume that the call initiated with station B. For every upcrossing when we have a previous downcrossing, we will have hand-off to station A; for every downcrossing when we have a previous upcrossing, we will have hand-off from A to B.

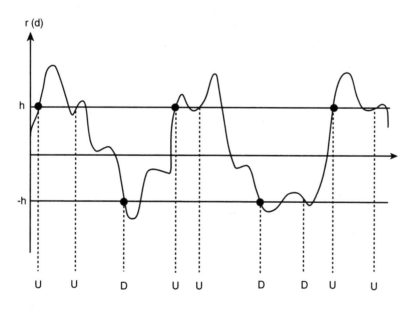

Figure 6.12 Hand-offs and Level Crossings

An important aspect of the hand-off process is the detection of the need of the mobile for a hand-off. A need for hand-off will result in forced termination if either the detection of that need is missed or the need for hand-off is detected but there is no channel available for hand-off.

Another important aspect that has not been discussed in this chapter is the signal outage. In such a case, if the mobile has already initiated a call, the call will be forced to terminate, or if it is in the process of hand-off, we will have hand-off failure. For the model, one should note that we should allow some number of crossings before we actually initiate the hand-off process so that we will avoid unnecessary hand-offs.

6.6 CONCLUSION

The process of hand-off is a very important aspect of cellular PCS communication. In this chapter, we have introduced the hand-off problem and analyzed some possible solutions to these problems, with the main concern being the field strength measurement. We showed that hand-off can be studied in terms of the level of crossings. Unfortunately, we could not obtain any measured data to confirm the results. On the other hand, it seems that the derivation of the exact model for hand-off will not be feasible because of the many factors that can be varying simultaneously.

Hand-off algorithms must be studied very carefully by the cellular systems designers; otherwise, the systems will lose customers. This is particularly true in Low Earth Orbit (LEO) satellite systems [7] where hand-offs must occur from satellite to satellite as they traverse their orbits.

Appendix to Chapter 6
MATLAB Code

Coding to estimate hand-off rate due to cell splitting:

Assume mobiles are distributed uniformly in the cell and direction of travel of mobiles (3 uniform distributed over [0,2 pi]).

```
G = 5
F = 0:1:1000;
H1 = G*F.^3;
plot(H1);
grid;
title('Effect of cell splitting on hand-off');
x label ('Factor of cell splitting');
y label (' # of hand-offs/cell");
pause;
```

Due to shape/orientation:

```
A = 100*100;
b = 0:1:1000;
H2=2*sqrt(A.*b);
plot (H2);
grid;
title(' Effect of cell shape (th=0)');
x label (' # of mobiles crossing the boundaries');
y label (' # of hand-offs/cell ');
pause;
```

Mean hand-off rate:

g = 0:1:1000;
m = 1.76;
H = (1/m). *g;
plot (H);
grid;
title (' Mean hand-off rate ');
x label (' Travel time ');
y label ('Average HO Rate');
pause;

REFERENCES

[1]   Yacoup, Michel D. *Foundations of Mobile Radio Engineering.* (Boca Raton: CRC Press, 1993).

      Mehrotra, Asha. *Cellular Radio Analog and Digital Systems.* Norwood, Mass: Artech House, Inc., 1994).

[2]   Lee, William C.Y. *Mobile Cellular Telecommunication Systems.* (New York: McGraw-Hill, 1989.)

[3]   Jean

      Asha, Mehrotra. *Cellular Radio Performance Engineering.* (Norwood, Mass: Artech House, 1994.)

[4]   Papoulis

      Linnartz, Jean-Paul. *Narrowband Land-Mobile Radio Networks.* (Norwood: Artech House, 1993).

[5]   Lecours, Michel. "Statistical Modeling of the Received Signal Envelope in a Mobile Radio Channel." *IEEE Transactions on Vehicular Technology*, Nov., 1988: 37.

[6]   Turin

[7]   Ricci, F.J., and V. Vaughn. "Residing Time Distributions and Call Handover Traffic Performance of Cells in LEO Mobile Satellite Systems." *Globecom 95.* Singapore, Nov., 1995.

[8]   Rappaport, Stephen S. "The Multiple-Call Hand-Off Problem in High Capacity Cellular Communication Systems." *IEEE Transactions on Vehicular Technology*, August, 1991: 40.

[9]   Hong, Daehyoung. "Traffic Model and Performance Analysis for Cellular Mobile Radio Telephone Systems with Prioritized and Nonprioritized Hand-off." *IEEE Transactions on Vehicular Technology*, August, 1986: 35.

[10] Ricci, F. Class Notes. Courie EE6644, Cellular Radio. Virginia Tech, Summer, 1994.

[11] Sanjiv, Nanda "Teletraffic Models For Urban and Suburban Microcells: Cell Sizes and Handoff Rates." AT&T Bell Laboratories. Holmdel, NJ.

Chapter 7

Wireless LANs

Wireless LANs are local networks, normally confined to a single building, that provide relatively high data connectivity to mobile terminals. The data rates can run from 64 kbps to 20 Mbps. Although wired networks can provide higher speeds, wireless LANs provide mobility and can potentially reduce installation costs, estimated to be as much as 40 percent of the overall installation cost of LAN. The great expansion of mobile computers, including notebooks and personal digital assistants (PDA), provides a large potential market for mobile LAN connectivity. The use of wireless LANs in a static environment has some advantages over cabled networks, such as flexibility and minimal cable installation; however, cabled networks still remain the dominant technique. Both radio frequency (RF)and infrared (IR) have distinct advantages in comparison with one another. These will be discussed subsequently.

Similar to wired versions, wireless LANs may use the same topologies—ring, bus and star. Several IR LANs have been developed using the ring architecture in which a token is passed from terminal to terminal according to Santamaria and Lopez-Hernarndez.[1] The bus architecture is emulated in wireless LANs by having all stations monitor a common channel by using a carrier sensing mechanism such as the one being promulgated in the new 802.11 standard. (See appendix to this chapter.) The star configuration works well when the flow of data is primarily from the center (star) node to the outlying nodes such as in a main-frame/remote terminal architecture.

This chapter provides a brief description and comparison of the two major technologies used in wireless LANs, infrared (IR) and radio frequency (RF) emitters. Emphasizing RF technology, the chapter then reviews some of the current developments in wireless LANS.

7.0 WIRELESS LAN TECHNOLOGY

RF technology uses narrow band schemes or spread spectrum. The FCC allocated the industrial, scientific and medical (ISM) bands (902–928 MHz, 2400–2483.5 MHz, and 5725–5850 MHz) for spread spectrum LANs. An additional 20 MHz is also available for wireless services in the 1910–1930 MHz band. According to [1], licenses are not required in any of the above bands if transmit power is under 1 watt, which limits transmission distances to approximately 800 feet. Spread spectrum techniques are preferred because of the potential reduction in multipath receive problems and because it potentially provides much higher bandwidths. Many spread spectrum devices are available on the market today.

Three major problems are detractors to the use of RF: frequency allocation, interference and security. Frequency allocation is limited for LANs, but since LANs operate with low power, frequency reuse is possible for the allocated bands. Also, other higher bands are being explored, according to [1]; for example, 17 and 61 GHz bands hold promise for larger bandwidths of over 100 Mbps. [2] Interference is a problem that can arise from other wireless LANs not under control of the same organization and from other industrial sources. The microwave oven, for example, operates in the 2.4 GHz band. This problem can be controlled by using spread spectrum techniques and by the required use of low transmit power for the modems. The problem may be solved by the use of the upper microwave and millimeter wave frequency bands, which offer a much less crowded spectrum. Because of the ability for the RF to penetrate walls, including the walls of the outside of a building, hostile operators can intercept LAN communications. This problem can be lessened with appropriate encryption technology.

IR uses both laser diodes and light-emitting diodes as emitters in the 840–950 nm wavelength. The chief problems encountered are those involving the emitter power, the multipath intersymbol interference, shadowing problems and the environmental noise. Because of eye safety considerations, laser diodes cannot be used inside but can be used for interbuilding communication links. LEDs, however, have a power-speed product problem since high-powered LEDs do not have fast switching times and fast LEDs (used for fiber optic communications) do not have the power required to diffuse the transmitted signal so that a line-of-sight (LOS) path is not required. In avoiding an LOS path, however, a multipath problem ensues because the use of the walls and ceiling provides a multitude of paths, which greatly lengthens the emitted pulse and slows down transmission rates. To lessen this problem, both active and passive "satellites" are used on the ceiling to retransmit or reflect, respectively, the emitted signal. (See Figure 7.1.) These satellites also lessen

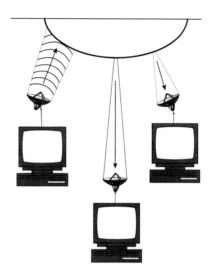

Figure 7.1 Passive and Active IR Satellites
Source: Reprinted from [1].

the problem of shadowing, in which a person in the wrong place may absorb the signal. Background light produces noise for the receiver and must be accounted for in any design.

Table 7.1 summarizes the comparison of IR and RF wireless LANs. IR is especially useful in high EMI environments such as are common in industrial environments. It is also inherently secure, since the IR signal cannot penetrate the walls. However, it is limited to a single room or to LOS links with satellites or with other terminals. On the other hand, RF provides the potential for truly mobile systems and is generally more practical for covering an entire building.

TABLE 7.1 Comparison of RF and IR Wireless LANs

Technology	RF	(Quasi) Diffuse/IR	LOS IR
Mobility	Good	Medium	None
Speed	.064–5.7 Mbps	1–20 Mbps	Very high
No. of channels	Large	One	One
Licensing	Restricted	None	None
Safety	Good (low pwr)	Good	Fair
Interference	Possible	None outside LAN	None
Security	Encryption req.	High	High if inside
Standards	802.11	IRDA 1.0	IRDA 1.0
Price	Med. to high	Med. to low	Med. to low

Source: Reprinted from [1].

This section of the chapter provides a sample of current developments in wireless LANs. Because of the scope of this chapter, the conclusions should be considered only preliminary until a more thorough market survey can be made. At the end of the section, Table 7.2 provides a listing of the vendors mentioned in the referenced sources and their stated activities in wireless LANs. (Reference [3] provides an extensive list of vendors supplying wireless equipment.) The current developments section is a compendium of articles on standards activity, announcements of new RF adapters cards, and applications of these adapter cards into portable computers. Table 7.3 also at the end of the section, summarizes the main characteristics of the LAN adapter cards.

Schwartz [4] indicates that the RF wireless LAN market is still small, since the vast majority of RF modems sold in 1993 (81,000) were for cellular (narrowband) connections. However, Edwards [5] appears very positive on market trends and quotes an estimate of $65 million spent on wireless LANs in 1994, double the amount for 1993, and predicts sales of $100 million in 1995. According to [3], an industrial base of 27 vendors compose the supplier infrastructure of IF and RF wireless LAN; AT&T Global Information Solutions is the current market leader. [5] LANs supplied by these vendors operate with speeds varying from 242 kbps to 20 Mbps over ranges from 80–800 ft, but real performance is typically one-half or one-third peak performance rates.

Two items highlight current standards work. Both IR and RF wireless LANs have had considerable activity in the standards arena. However, the IR standard is limited in scope (low transmission rates). Six items indicate new or current products on wireless LAN adapter cards. All RF products reviewed, except one, use spread spectrum techniques in the 2.4-GHz band. Effective throughput rates advertised vary from 1.0–5.7 Mbps; transmission range varies for different devices and environments. Portable PCs with PCMCIA adapters or desktops using RF modems inserted in the bus (ISA or Micro Channel) or connected with a regular cable adapter can connect with one another as peers or to servers through a bridge, called an access point. One item discusses IBM's market entry, which indicates that the technology is maturing. IBM's offering differs from that of others because it offers the same hardware for both the access point (or base station) and the stations. Three items on wireless LAN applications indicate a trend for using the mobility of the wireless LANs to enhance efficiency of retail operations. All three mention small mobile computers—either personal digital assistants or notebooks.

7.1 STANDARDIZATION EFFORTS

The productivity of wireless LANs has led to a number of standards which are discussed below.

7.1.1 IEEE 802.11 Wireless LAN Standard

In November 1993, the IEEE 802.11 committee voted to select the Distributed Foundation Wireless Media Access Control (DFWMAC) as the foundation for the 802.11 media access control (MAC) specification. To gain broader recognition, the committee is coordinating the standard with other groups such as the standards groups within the Telecommunications Industry Association and the European Telecommunications Standards Group (ETSI). The protocol features a distributed access control mechanism similar to Ethernet's, in which each node senses the carrier (Carrier-Sense Multiple Access/Collision Avoidance). It also features power management, synchronization and support for time-bounded services such as voice and process control. Minimum bandwidth is set at 1 Mbps. [6] (See the appendix to this chapter for more detail).

7.1.2 Ratified IR Device Specification For Wireless LANs

The Infra-Red Data Association (IRDA), a consortium of 75 vendors, announced in July 1994 the formal ratification of the IRDA 1.0 protocol. The protocol standardizes IR connections between vendors' portable computers and personal digital assistants to LANs and network-based peripherals; it uses the technology developed for television controls. The IRDA-compliant hardware components are expected to cost only several dollars, so that vendors could incorporate them with minimal cost impact. The standard allows a 3-meter, 30-degree field of view and a 115.2 kbps rate. Notebook PCs and peripheral devices with the new standard are expected this fall from several vendors. Microsoft is developing an IRDA driver in its next version of Windows™. [7]

7.2 WIRELESS LAN ADAPTERS

There are numerous innovative LAN devices and adapters in the market today. American Airlines is utilizing a hand held spread spectrum device to track cargo at airports around the world. Included below are some of the devices available.

7.2.1 IBM Wireless LAN Adapters

IBM is currently shipping its first wireless LAN adapters, after receiving final FCC approval in June 1994. IBM did not wait for the IEEE 802.11 standard but will comply when the standard is completed. IBM chose a TDMA protocol (vice the normal carrier sense multiple access) that can distribute wireless traffic to up to 50 users in each of 20 cells. Each cell contains a nondedicated base station that can also be connected to a wired LAN. IBM's product line differs from others in that any terminal with the adapter can be the base station instead of requiring special hardware for the base station. Base stations run software to monitor the links in order to modify the cells and to modify the frequency hopping patterns to lessen interference, using a proprietary "French protocol." Cells can overlap and can extend in radius in open air to 600 feet to the remote terminal. [8, 9, 10]

IBM offers the LAN adapters, priced at $795, with a PCMCIA interface for mobile computers or ISA or Micro Channel bus interference for desktops. Each adapter is equipped with an antenna and spread spectrum transceiver and can be used as a remote terminal or the base station. "DES-like" encryption is an option. Software to bridge the gap between a wireless LAN and a wired LAN (Ethernet or token ring) is available for the base station. The LAN uses the 2.4-GHz band at a 1 Mbps data rate. Overhead from coding is offset by data compression to maintain throughput. The product line will also have WAN adapters for the Ardis wireless network and voice, data, fax and paging services to NetWare® and OS/2® based Ethernet or Token Ring LANs.

7.2.2 Proxim Announces New LAN Adapter

Proxim will be supplying its newest wireless LAN adapter, RangeLAN2™ to Zenith Data Systems for its notebooks. The RangeLAN2 operates at 1.6 Mbps across 15 channels in the 2.4-GHz band and a maximum range of 1,000 feet in open areas. Proxim also has previously assisted other manufacturers, Grid Systems Corp. and Fujitsu Personal Systems Inc., with its RangeLAN transceiver, which ran at 242 Kbps. It also had provided support to Compaq Computer Corp.'s notebooks. The RangeLAN2 products will begin shipping in August 1994 and are listed at $695 for the PCMCIA adapter, $595 for the ISA-bus version, and $1,895 for the access point (base station). [9, 11]

7.2.3 Solectek's Wireless Ethernet Adapter

Solectek started shipping in July 1994 a battery powered RF Ethernet adapter, called AirLAN/Parallel™. It is designed for portable computers and connects to their parallel ports. The device uses spread-spectrum techniques in the 902–928 MHz band and is rated for a range of 800 feet for a data rate up to 2 Mbps. The rate is constrained, however, by the parallel port, which is currently limited to 1 Mbps. The 10-ounce, $699 device can connect to an AirLAN/Hub™ or a PC base station equipped with an AirLAN/Internet™ adapter. The batteries are rechargeable and are rated for 10 hours of operation. [12]

7.2.4 Xircom, Inc. Ships New LAN Cards

Xircom began shipping in July 1994 its Netwave™ wireless LAN products, consisting of a PCMCIA card with integrated antenna and an Ethernet bridge. The use spread spectrum techniques is in the 2.4-GHz band with effective throughput of about 1 Mbps data rate. The Ethernet bridge, which serves as the base station for access to the LAN server, can connect terminals up to 150 feet away. According to current plans, these products will be enhanced with data compression, encryption option, improved power management and incorporation of the Simple Network Management Protocol (SNMP) for the Ethernet bridges. Listed costs are $599 for the adapter and $1,499 for the access point (bridge). [9]

7.2.5 Dayna Communications Adapters for Macintoshes

Dayna Communications Inc. plans to start shipping in October 1994 the Roamer™ family of wireless LAN adapters that are compatible with the Apple® MacIntosh®, PowerBook™ and Newton computers. Roamer is based on the technology of Xircom and transmits data up to 1 Mbps at a maximum range of 150 feet in the 2.4-GHz band. Roamer supports AppleTalk®, Mac® TCP, and Mac IPX protocols. It is very lightweight: 4.5 ounces. [13]

· 7.2.6 Status on AT&T Global Information and Motorola LAN Adapters

AT&T Global Information Solutions is the current market leader in RF wireless adapters for LANs and uses direct sequence spread spectrum technology with its WaveLAN™ products. Throughput rate is advertised at 2 Mbps at ranges from 10 to 180 meters, depending on the environment. AT&T also offers the WavePoint™, which is an access point for the integration of wired and wireless LANs; it uses an omnidirectional antenna. [5]

Motorola offers Altair Plus™ II as its third generation product for wireless LANs. These products attach directly to existing Ethernet cable adapters and therefore can connect anything that supports Ethernet. Throughput rates are listed as 5.7 Mbps. Altair Plus II includes SNMP network management and features a control module to operate with 50 user modules in a work area of up to 50,000 square feet. The control module can connect to an existing wired Ethernet or connect directly to a file server to set up an independent wireless network.

7.3 WIRELESS LAN APPLICATIONS

7.3.1 K-MART

K-Mart plans to equip all of its store managers in its 2,431 stores with a hand-held computer that has a built-in RF modem and bar-code scanner. The device is the Symbol™ PPT4100, announced by Symbol Technologies Inc. The device has a pen-based user interface, used to touch icons that download pricing or inventory information from the local or corporate databases after a product bar code is swiped over its bar-code reader. The device will also be used to send and to receive messages from the manager to the corporate headquarters. [14]

The RF modem is based on spread spectrum technology over 908–926 MHz and is connected to the local token-ring LAN via a base station co-located with a PC. These token ring LANs are connected to the corporate databases at Troy, Michigan, via a satellite X.25 based network. K-Mart planned to have two to five of the PPT4100 personal digital assistants for each store by Spring 1995. The PPT4100 also has the capability to support RF modems that can connect to the RAM Mobile Data Inc., Mobitex network, the Ardis network or cellular digital packet data network via two PCMCIA slots.

7.3.2 Chicago Board of Trade (CBOT)

The Chicago Board of Trade announced that by the end of 1995, 100 of its traders should have hand-held computers to replace the paperwork used to record transactions. It will allow near real-time posting of transactions instead of the next-day time frame experienced currently. If the system works, all 3000 traders would have hand-held computers by the end of 1995, and CBOT could be the first major trading floor in the U.S. to fully employ wireless technology. The Board has already spent $11 million on the system; and the total is expected to be $30 million upon completion. Made by Seiko, the hand-held computers have been tested over the past year; they use an RF modem by Proxim.

On the trade floor, CBOT will install about four cells, which will be connected via fiber to an Ungermann-Bass router. The route will connect the cells to a Sun® SPARCserver™ 1000 running a Sybase® SQL Server™ 10. Transactions are confirmed by a parallel Sun SPARCserver, which posts them to the mainframe. [15, 16]

7.3.3 Zenith Data System (ZDS)

In collaboration with Proxim, ZDS announced that it is developing a hand-held PC with an integrated LAN link. ZDS already had Ethernet and client "shells" built into their notebooks, but these were constrained by wire lines. ZDS units will be using Proxim's RangeLAN2 (see Proxim entry). The plan is to have it released by year's end, 1994, although it is dependent upon Microsoft's WinpPad™ operating system, which is also not expected until the year's end, 1994. [8, 15]

Listed in Tables 7.2 and 7.3 are examples of vendors and LAN adapters that are currently evolving.

Aironet has wireless radios and LAN devices at 900 MHZ and 2.4 Ghz using spread sprectrum techniques.

TABLE 7.2 Vendor List

AMD (802.11 std.)

AT&T Global Information Solutions/NCR Microelectronic (direct Sequence Spread Spectrum Modems, 802.11 std)

Dayna Communications Inc., Salt Lake City, UT, 1-800-443-2962. (Mobile Access Server, MAC LAN adapters)

DEC (802.11 std)

IBM (RF modems/base stations)

International Computers (802.11 std)

Microsoft (IRDA software drivers)

Motorola (RF LAN adapters)

National Semiconductor (802.11 std)

Norand (802.11 std)

Proxim Inc., Mountain View, CA. (RF modems)

Solectek, San Diego, CA, 619-450-1220. (RF modems/base stations)

Spry, Seattle WA, 1-800-777-9638. (Air series of software)

Symbol Technologies, (Personal data assistants with built-in RF modems)

Telxon (802.11 std)

Xircom, Calabasas, CA (Netwave RF modems/base stations)

Zenith Data Systems, 1-800-533-0331. (RF-equipped notebook, computers with Cruise LAN)

TABLE 7.3 Summary of Selected LAN Adapters

	IBM	Proxim	Solectek	Xircom	Dayna[1]	AT&T	Motor
Freq	2.4 GHz	2.4 GHz	902 GHz	2.4 GHz	2.4 GHz	2.4 GHz	2.4 GHz
Range (Max)	600 ft	1000 ft	800 ft	150 ft	150 ft	600 ft	unk
Throughput	1 Mbps	1.6 Mbps	2 Mbps	1 Mbps	1 Mbps	2 Mbps	5.7 Mbps
Access	TDMC	unk	unk	CSMA(?)	CSMA(?)	unk	unk
Price (Sta)	$795	$695/595[2]	$699	$599	unk	unk	unk
Price (AP)	N/A	$1,895	unk	$1,499	unk	unk	unk
Encryption	yes	unk	unk	yes	yes	unk	unk
Product	unk	RangeLAN2	AirLAN	unk	Roamer	WaveLAN	AltairP

[1]Dayna uses Xircom technology   Unk = Unknown
[2] PCMCIA/ISA-bus prices   CSMA = Carrier sense multiple access
Motor = Motorola

7.4 WIRELESS LAN APPLICATIONS

Although many products for RF and IR wireless LANs are available, the introduction of applications is just beginning. Presently, research and trials are being conducted for wideband wireless LANs (2–20 Mb/s). Some research considers the use of asynchronous transfer mode (ATM) protocol for laptop communication to portable base stations. [17] Figure 7.2 presents an overview of the proposed architecture.

Typical mobile users are assumed to be laptop or notebook computers. Services supported include conventional data applications (e.g., over TCP/IP or SPX/IX) as well as multimedia (video, voice and data) applications.

The proposed LAN consists of network nodes called portable base stations (PBS) providing microcell coverage. The PBSs are designed to be low cost, compact and high speed and can be relocated conveniently. The links between PBXs are optical loser links. The user-to-PBX access links are primarily for mobile access (e.g., 2–20 MHz) and are wireless RF or IR links. There are many routing, flow control and error-control challenges in ATM networks. For example, with mobile users, what happens to the ATM virtual path (VP) and virtual circuit (VC) concept? The mobile network concept discussed in Chapter 8 looks at some of the protocol problems.

The wireless LAN concept outlined in Figure 7.2 could also be configured by using packets (x.25) over RF or IR links. However, in this case the data rates achievable would be much lower. The base stations in this case could be Private Branch Exchanges (PBXs) for voice and lower speed data (2.4 kb/s–sMb/s). Pocket radios (x.25) could be incorporated with laptop computers for communications to PBS/PBS. In this case, looser links would not have to be used. Perhaps the Unlicensed Personal Communications System (UPCS) band could be use for transmission.

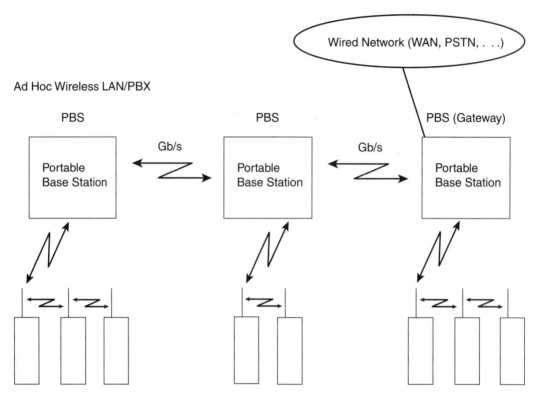

Figure 7.2 A Wireless Ad Hoc Network System Architecture

Additional research is needed to prove out both the ATM and packet concept of a wireless LAN. However, both seem to have considerable potential for wideband data communication.

7.5 CONCLUSIONS

Wireless LAN systems and implementation are beginning to emerge as an exciting technological advancement. Many products are already available for use in the marketplace, and research is being performed for wideband data applications. The unlicensed band seems to have good potential for utilization in the application. Generally, the application of technology to address this area offers opportunities for new business and research.

Appendix to Chapter 7
Distributed Foundation Wireless Media Access Control (DFWMAC)

The DFWMAC protocol supported by the IEEE 802.11 committee is designed to support file transfers, program loadings, transaction processing, multimedia and manufacturing process control at the 1 Mbps rate. [6] The variance in the timely delivery of packets is minimized so time-bounded services (such as voice) can be supported. The protocol was written to support the needs of a variety of users, such as financial, retail, office, school and industrial users. Mobile computing at the rate for pedestrians or optionally for vehicles can also be supported. The wireless LAN architectures supported by the protocol divide into two main types: ad hoc and infrastructure based. The ad hoc LANs are those that are set up instantly by any number of users for temporary purposes, such as a short-term project or a conference. The infrastructure-based architecture is one in which mobile users can roam through a building while maintaining contact with the organization's main computer resources through an infrastructure that is usually wired.

The different components of a wireless LAN are defined in Figure A.1. A cell within an infrastructure-based network is known as a basic service area (BSA), which can contain any number of discrete groups of wireless stations. Multiple BSAs can be connected by access points (AP) to form an extended service area through a distribution system (which is normally wired). The set of stations serviced by a single AP is known as a basic service set (BSS); the collection of BSSs connected through the distribution system is known as the extended service set (ESS).

REFERENCE MODEL

The 802.11 protocol covers the OSI physical and link layers. The physical layer is specified by a series of different physical specifications covering a number of different fre-

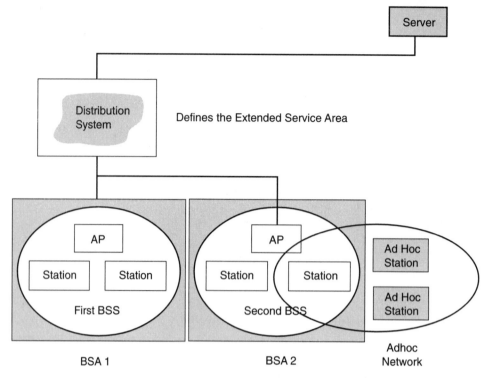

Figure A.1 Architecture for Wireless LAN
Source: Reprinted from [9]

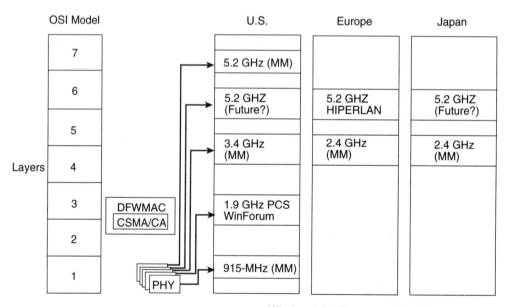

Figure A.2 Reference Model for 802.11

quency spectrums and ID. (See Figure A.2) The link layer portion, defined for media access control by the Distributed Foundation Wireless Media Access Control (DFWMAC) protocol, has a specification common to all wireless networks.

ACCESS MECHANISM

Using the Distributed Coordination Function (DCF) protocol, DFWMAC provides the Carrier-Sense Multiple Access/Collision Avoidance (CSMA/CA) mechanism with acknowledgment for asynchronous transfers between stations. The station determines if signal energy above a certain threshold exists; if it does not, it is considered available and the station transmits a frame of data. The station waits for an acknowledgment from the distant station to send another frame, or it resends the original frame after a timeout period. If not, the channel is considered busy and the station waits a random amount of time, using an exponential "backoff" algorithm. This algorithm has three priorities, in which the back-off times are random within each priority. The priorities are:

- SIFS (short interframe space): The shortest wait time, which is reserved for immediate response actions such as frame acknowledgments, request to send (RTS), clear to send (CTS), and any contention-free response frame sent during time-bounded services.
- PIFS (point coordination function IFS): Intermediate time, which is used by time-bounded services to send frames.
- DIFS (distributed coordination function IFS): Longest time, which is used by the asynchronous frames during contention.

Using the above mechanism, the MAC layer is able to provide reliable communications on a frame-to-frame basis.

OTHER FEATURES

DFWMAC provides power management and time-bounded services. For power management, it allows power conservation of battery-operated portables by having a protocol that switches stations from full-power mode to low-power or sleep mode. For time-bounded services, it also has an optional Point Coordination Function (PCF) protocol, which runs on top of the basic CSMA/CA protocol to provide contention-free service. It does this by defining a superframe to ensure contention-free service. When a station needs to transmit PCF, it senses whether the medium is free. If it is, the station starts sending PCF data for a certain contention-free period, since its interframe intervals (PIFS/SIFS) block out the longer wait times for the asynchronous frames (DIFS). The superframe can vary in length because after the contention-free period, a contention period ensues and the PCF protocol will not interrupt a frame in transmission to start the beginning of another superframe and, consequently, another contention-free period. See Figure A.3.

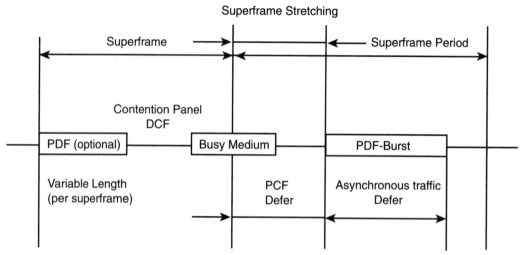

Figure A.3 The Superframe
Source: Reprinted from [9]

REFERENCES

[1] Santamaria A., and F. Lopez-Hernandez. *Wireless LAN Systems*, pp. 1–20, 129–243, Norwood, MA, Arctech House, 1994.

[2] Fernandes, J., P. Watson, and J. Neves. "Wireless LANs: Physical Properties of Infra-Red Systems vs. Mmw Systems." *IEEE Communications Magazine*, Aug., 1994: 32(8):68–73.

[3] "Wireless Equipment Directory." *Communications News*, Aug., 1994: 31(8):40–46.

[4] Schwartz, J. "Study Says Wireless-Modem Shipments to Increase Substantially By 1998." *Communication Week*, June 13, 1994: (509):43.

[5] Edwards, M. "Surge In Popularity Spawns New Ideas For Wireless LANs." *Communications News*, Aug., 1994: 31(8)40–46.

[6] Links, C., W. Deipstraten, and V. Hayes. "Universal Wireless LANs." *Byte*, May, 1994: 19(5):99–108.

[7] Schwartz, J. "Vendor Group Ratifies Infrared-Device Spec." *Communication Week*, July 25, 1994: (515):33, 36.

[8] Dryden, P. "IBM Set With New Wireless LAN Adapters." *Communications Week*, Jun 20, 1994: (510):75.

[8] Jenks, A. "Wireless Hardware Just Now Arriving." *Washington Technology*, Jan 13, 1994: 8(19):28.

[9] Dryden, P. "Wireless Cards Help Create Mobile LANs." *Communications Week*, July 25, 1994: (515):21–22.

[10] "IBM Readies Wireless LAN Line," *Info World*, July 18, 1994: 16(29):54.

[11] Dryden, P. "Wireless LAN Link to Come From ZDS." *Communications Week*, May 16, 1994: (505):96.

[12] "AIRLAN/Parallel Offers Wireless Ethernet Access." *Info World,* July 11, 1994: 16 (28):52.

[13] "Wireless Links to Ship For Macs, Portables." *Info World*, Aug. 8, 1994: 16(32):38.

[14] Schwartz, J. "K-Mart to Use Pen-Based Computing." *Communication Week*, Jan. 17, 1994: (488):27.

[15] "Manager's Bulletin Board." *Info World*, July 18, 1994: 16(29):58.

[16] Schwartz, J. "Trading Floor Set to Go Wireless." *Communication Week*, May 30, 1994: (507):8.

[17] "Bahama: A Broadband Ad-Hoc Wireless ATM Local Area Network." *K.Y. Eng.*, ICC'95, Seattle, Washington.

Chapter 8

Adaptive Mobile Network

In order to provide an example of emerging PCS designs, a unique network is considered. The Adaptive Mobile Network is designed to provide a full-duplex, wireless, packet-switched network to mobile terminals. Such a network could support a wide variety of applications, such as delivering directions or maps to drivers based on their GPS position or allowing traveling children to play games with children in other cars. The topology of the system is modeled after the existing AMPS cellular phone networks. This system differs from AMPS, however, as it is a packet-driven network using the concept of data channels. The data channels are analogous to existing Ethernet channels. Operationally, mobile units do not exchange packets directly; all traffic is routed through the cell site. The mobile units transmit packets to a cell site, where they are either processed or routed to their destinations. All message traffic destined for a particular mobile unit is delivered by the cell site where the mobile unit currently resides. In addition to the physical connections and operations, a control structure has been designed and simulated to keep track of the location of the mobile units, and frame structures are presented that integrate the network into a seamless end-to-end delivery system. The design discussed in this chapter utilizes a concept of high data rate packets.

8.0 THE ADAPTIVE MOBILE NETWORK

The Adaptive Mobile Network was developed to conform to the Open Systems Interconnection (OSI) model [1] for computer communications, as depicted in Figure 8.1.

The seven layers of the OSI model are defined Table 8.1. This design effort addresses the physical, data link, network and transport layers of the model individually. The intent is to design a reliable packet network supporting multiple applications developed by many vendors. Because they are application dependent, the session, presentation and application layers are out of the scope of this project and will be designed at a later time.

8.1 DESIGN

The design is divided into four parts corresponding to the four lowest levels of the OSI network model. The physical design section describes the topology of the network and the modulation technique used for the forward and reverse channels. Link budget calculations and a UHF frequency sharing scheme are also presented here. The data link design section

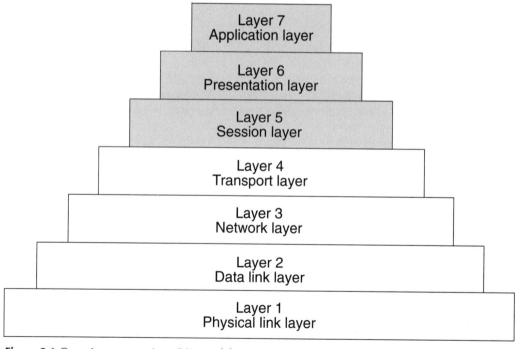

Figure 8.1 Open Interconnection (OSI) Model

Table 8.1 The OSI Layers

1	Physical	Concerned with transmission of unstructured bit streams over physical medium; deals with the mechanical, electrical, functional and procedural characteristics to access the physical medium
2	Data link	Provides for the reliable transfer of information across the physical link; sends blocks of data with the necessary synchronization, error control and flow control
3	Network	Provides upper layers with the independence from the data transmission and switching technologies used to connect systems; responsible for establishing, maintaining and terminating connections
4	Transport	Provides reliable, transparent transfer of data between end points; provides end-to-end error recovery and flow control
5	Session	Provides the control structure for communication between applications; establishes, manages and terminates connections (sessions) between cooperating applications
6	Presentation	Provides independence to the application processes from differences in data representation (syntax)
7	Application	Provides access to the OSI environment for users and also provides distributed information services

describes the protocols used for data transfer across the physical links. The network design section describes how hand-offs will be used to make the design appear to the users to be a seamless network. Also included in this section is a discussion of how users join and exit the network. The final layer designed for this project, the transport layer, describes how addressing will be accomplished for end-to-end connections.

8.1.1 LAYER 1 PHYSICAL DESIGN

The goal of the Adaptive Mobile Network is to provide mobile users access to a wireless packet network capable of full duplex data exchanges. The main challenge for the physical design is to provide a highly accessible, reliable service while using as little bandwidth as possible. To accomplish this, the network uses a frequency sharing scheme similar to the existing cellular telephone network. For clarity, this section is broken into four parts; cell design, cellular topology, frequency allocations, and power calculations.

Cell Design

Within each cell, there is a cell-to-mobile (C-M) channel and a mobile-to-cell (M-C) channel. The C-M channel is a constant bit rate channel that the cell uses to transmit to all mobile units currently in the cell's footprint. Each mobile unit must constantly monitor this channel, searching for data addressed to it. Table 8.2 summarizes the channel specifications.

Table 8.2 Summary of M-C and C-M Channels

	Forward Channel (C-M)	Reverse Channel (M-C)
Modulation	DQPSK	DQPSK
Bandwidth	1.6 MHz	3.4 MHz
Es/No for 10-6 Symbol error rate	15.5 dB	15.5 dB
Data Rate	1.544 Mbps	3.088 Mbps

The C-M channel data rate of 1.544 Mbps (T1 rate) is provided by using differential quadrature phase shift keying (DQPSK) in a 1.6-MHz channel. The required SNR to achieve a 10^{-6} symbol error rate is 15.5 dB [2]. A DQPSK transmitter and coherent receiver are shown in Figure 8.2 and Figure 8.3. In addition to the low signal to noise ratio, DQPSK modulation was chosen because it allows the mobile units to recover the clock from the waveform and is resistant to channel inversions.

As in the C-M channel, the M-C channel data rate of 3.088 Mbps is provided by using Differential QPSK in a 3.4-MHz channel. The M-C channel is twice as wide as the C-M channel to reduce congestion problems (discussed in the data link design section). The M-C transmitter and receiver are functionally identical to the C-M channels except that the transmitter is controlled by a device that detects when the bandwidth is available.

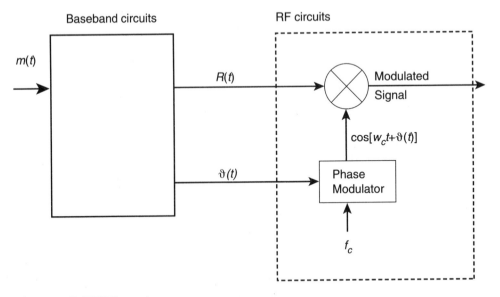

Figure 8.2 DQPSK Transmitter

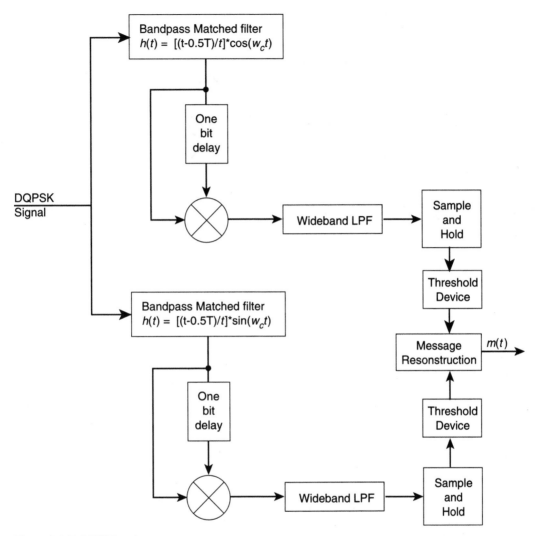

Figure 8.3 DQPSK Receiver

Cell Relations

As in the existing AMPS cellular system, frequency sharing is accomplished through a repeating building block consisting of seven cells. Each cell is designed to use a 5 MHz spectrum; resulting in a 35 MHz bandwidth for the entire system. Figure 8.4 shows the spatial configuration of the seven cells within the block and how the blocks fit together.

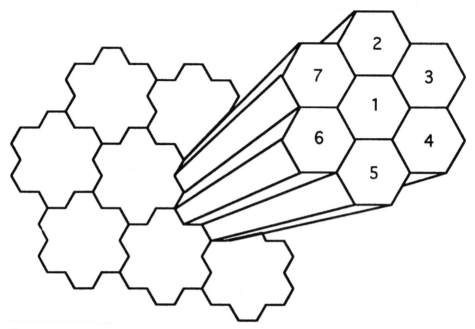

Figure 8.4 Frequency Reuse Pattern

8.1.2 LAYER 2 DATA LINK DESIGN

Frames are sent across both the forward and the reverse channels, using the ISO standardized high-level data link control (HDLC). Figure 8.5 shows the HDLC frame format. Stallings [3] provides a detailed discussion of HDLC.

The flag fields indicate the start and end of each frame. Use of the address field is explained in detail in the layer 3 discussion. Briefly, the address field determines the destination of the frame. The control field contains a busy bit that the cell will use to indicate to mobile units that it is ready to receive frames. The information field is used by the network layer to exchange user data, and the frame check sequence (FCS) field provides error detection.

Operationally, HDLC permits transmission of three types of frames (the control field indicates the type of frame being sent). The most common frames are information frames (I-

Flag	Address	Control	Information	FCS	Flag

| Bits: | 8 | 16 or 32 | 8 or 16 | Variable | 16 or 32 | 8 |

Figure 8.5 HDLC Frame Format

frames), which are used to exchange user data. Supervisory frames (S-frames) are used for flow and error control. Unnumbered frames (U-frames) are used for control functions such as cell hand-offs. The frame type is indicated in the control field, as depicted in Table 8.3.

The M-C channel is used by all mobile units to transmit frames to the cell site. This channel is analogous to an Ethernet link in that only one mobile unit within a cell at a time can communicate over this channel. The cell must indicate the status of the M-C channel by setting the busy bit in the control portion of the frames it transmits over the C-M channel. In the event the cell has no frames to transmit, it will send S-frames to indicate changes in the status of the M-C channel. The mobile unit must wait for the cell's ready indication before transmitting any frames. If two mobile units transmit at a given ready indication, a frame collision occurs and the cell will not acknowledge receipt of either frame, thus requiring the mobile units to retransmit damaged frames. For this scheme to work, a relatively wideband channel is required for transmission of relatively small packets. Otherwise, users could experience long delays for transmission, or even an inability to find a silent spot to insert their packet. This is the justification for providing the M-C channel twice the bandwidth of the C-M channel.

8.1.3 Layer 3 Network Design

This section describes in detail the network operations and addressing schemes—how the lower levels are integrated into a functioning network. First, the network topology and the mobile unit's addresses are described. This is followed by a description of how mobile units join and exit the network and finally, the hand-off technique is described.

Network topology

As described before, the network is constructed of geographically distributed cells. From a networking point of view, each cell can be viewed as a LAN. As depicted in Figure 8.6, groups of cells are connected to a central switching office (CSO) via landlines to form a MAN. Finally, the central switching offices are interconnected to form a WAN, which allows routing of frames across geographically distant points. In addition to providing access to distant terminals, the CSO is also where large, application-specific databases (such as maps) reside. Although physical and data link layers for the landline circuits are beyond the scope of this project (they are generally leased circuits), it is important to present the connections at the network level in order to show the long-haul capabilities of the network.

Table 8.3 Control Field

bit 1	bit 2	Busy bit (bit 3)	bits 4 to 8	Frame type
0	0 or 1	0 or 1	Frame sequence number	Information
1	0	0 or 1	Supervisory functions	Supervisory
1	1	0 or 1	Unnumbered functions	Unnumbered

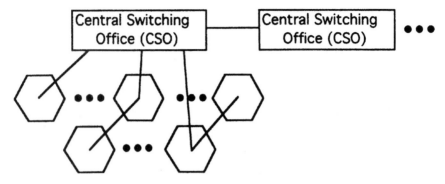

Figure 8.6 Network Topology (Provided via Landlines)

Addressing

Each mobile unit is assigned a 32-bit unique identification number (UIN) when it is constructed. This number is used by the network to differentiate among the mobile units. Once the mobile unit is a member of a cell, it is assigned a shortened local address by the cell. The cell keeps adjacent cells and the CSO informed of all the members of it's LAN.

Connecting

Before a mobile user can request or receive frames, it must first join a cell. This is analogous to attaching an Ethernet connection to your PC. When turned on-line, the mobile unit scans all possible C-M channels to determine which one has the strongest signal. Once this determination is made, the mobile unit attempts to join that cell by sending a U-frame across the corresponding M-C channel. This frame is a request to join the cell and indicates the mobile unit's UIN. Upon receipt of the request, the cell informs adjacent cells and the CSO of its new member and returns an acknowledgment that assigns the mobile unit a shortened 16-bit address for use while in the cell. At this point the mobile unit is permitted to broadcast I-frames into the M-C channel according to the rules defined in the data link section. Before going off-line, the mobile unit must inform the cell that it is no longer available to receive frames. The cell will pass this information along to all adjacent cells and to the CSO.

Hand-off

The mobile unit is responsible for determining the need for and executing a hand-off to another cell. For clarity, the hand-off process is pictured in Figure 8.7. While on-line, the mobile unit periodically monitors the signal strength of the C-M channels in adjacent cells. If another cell is significantly stronger than the current one, the mobile unit sends a join request into the new cell (step 1 in Figure 8.7). As before, the new cell sends an acknowledgment and informs adjacent cells and the CSO of its new member (steps 2 and 3 in Figure 8.7). The old cell receives this notification and realizes that the mobile unit is no longer under its footprint. Any remaining frames destined for that mobile unit will be forwarded to the new cell. The old cell, in turn, informs the previous cell (if any) that the

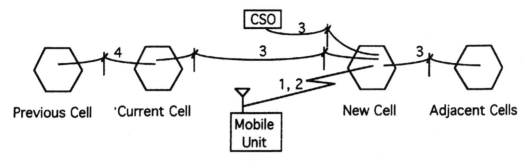

Sequence	Channel	Description
1	M-C	Request to Join
2	C-M	Acknowledgement
3	Landline	Membership announcement
4	Landline	Release of forwarding responsibilities

Figure 8.7 Hand-off Sequence of Events

mobile unit has been handed off (step 4 in Figure 8.7). This releases the previous cell from any forwarding responsibilities and allows the cell to reassign the local address.

The complexity of the handoff operation requires significant address management functions both at the cell and the CSO. In this regard, the appendix to this chapter provides a detailed explanation of a program that simulates the address management at the cell site.

8.1.4 Layer 4 Transport Design

The goal of the transport layer is to provide reliable, implementation-independent transfer of data between endpoints for the higher levels. The best way to explain how this is accomplished is to trace the transfer of a message from injection into the network until it reaches the destination.

At the mobile unit, the transport layer is responsible for accepting data (addresses and information) from higher levels and assembling it into data link frames for transmission. This includes generation of the FCS (frame check sequence) for error control. The frame is transmitted to the cell site by the lower layers and passed back up to the transport layer. At the cell site, the transport layer examines the destination UIN to determine where to send the frame. If the cell lists the UIN in its databases (which list all members of the cell and each adjacent cell), the cell will deliver the message to the appropriate cell for transmission to the recipient. If the UIN is not in the databases, the transport layer passes the frame to the CSO. In either case, an acknowledgment, indicating receipt of the frame, is returned to the sending mobile unit. At the CSO, the transport layer again searches the local database for the UIN and either passes the frame to the destination cell or to another CSO as required. Each time the frame is passed across a link of the network, an acknowledgment is sent to the previous node confirming receipt. Eventually, the frame arrives at the destination cell. At this point the transport layer replaces the UIN with the local address and delivers the frame. Within the recipient, the transport layer collects all the frames associated with this message and assembles the message for the higher levels.

8.2 CONCLUSION

The Adaptive Mobile Network presented herein is capable of providing full-duplex, wireless packet switching to mobile units. To date, the mobile links and the network operations have been designed. Design issues that have not been addressed include the MAN and WAN landline physical and data link layers, and the Application, Presentation, and Session layers of the network. Additionally, before a system could be deployed, a scheme for billing customers would need to be developed.

Appendix to Chapter 8
Database Requirements

The cell site is responsible for maintaining the following databases:

Unused Local Addresses
16 bit addresses 65536 Possibilities

Status of In-Use Addresses		
Local Address	UIN	Status
16 bit	32 bit 4,294,967,296	• In Cell • Forward to ... • Not Responding

Adjacent Cell Membership	
Cell Identification	UIN
• One of six	• List of all active Members in Cell

The CSO is responsible for maintaining the following databases:

Mobile Units Represented	
UIN	Cell ID
32 bit	Up to 42

The mobile units will maintain application-specific databases, such as mailing lists or lists of players in a game.

REFERENCES

[1] Couch, Leon W. *Digital and Analog Communication Systems.*, 4th ed., (location Macmillan, 1994).

[2] Ziemer, Rodger E., and Roger L. Peterson. *Introduction to Digital Communication.*, location Macmillan, 1992).

[3] Stallings, William. *Data and Computer Communications.*, 4th ed., (location Macmillan, 1993).

[4] Lee, William C. Y. *Mobile Cellular Telecommunication Systems.*, (location McGraw Hill, 1989).

Chapter 9

Speech Encoding of Voice in Digital PCS Systems

Cellular networks in metropolitan areas are becoming saturated: too many users for the given number of channels. The current analog systems, i.e., AMPS, ETACS, and JTACS, are not capable of handling the ever-increasing number of cellular phone users. The most widely held solution has been to turn toward digital cellular networks, which are capable of servicing more users than the analog systems. In fact, many digital networks in place in both North America and Europe improve upon the analog system threefold. It is already foreseen that in as few as 5 years these digital networks will also be saturated and new improvement schemes will be necessary. At the heart of these digital cellular networks is the digital encoding of the voice signal. It is this speech encoding that this chapter discusses.

This chapter initially discusses the speech signal and particular elements of speech that are of concern to modern speech encoding routines. Then, the chapter focuses on several basic approaches to speech encoding. A brief discussion of the analog cellular system used in North America (AMPS) and a discussion of the current digital cellular networks, both implemented and proposed, follows. Next, the chapter focuses on describing the current speech encoding schemes used in today's digital cellular networks. And finally, more exotic speech encoding techniques necessary for further increasing the capacity of cellular networks are discussed.

9.0 THE SPEECH WAVEFORM

Before we begin to discuss the features of speech, a short overview of the speech waveform is necessary. Speech is simply the acoustic wave that is radiated from the body's vocal system. When air is expelled from the lungs, the resulting air flow is perturbed by a constriction somewhere along the vocal tract. [2] This allows us to make different sounds.

The body's vocal system is made up of the vocal tract and the nasal tract. The vocal tract begins at the vocal cords and ends at the lips and consists of the pharynx and oral cavity. In the typical male, the length of the vocal tract is about 17 cm. The cross-sectional area of the vocal tract, as defined by the position of the tongue, lips, jaw and velum varies from 0 cm^2 to about 20 cm^2. The nasal tract becomes acoustically coupled to the vocal tract when the velum is lowered, which produces the nasal sounds of speech. As sound propagates down these tracts or tubes, the output frequency spectrum is shaped by the frequency selectivity of the tube. The vocal and nasal tubes act as an instrument much like those of a pipe organ or flute. The resonant frequencies of the vocal tract are called formants. These formant frequencies are regulated by the shape and dimensions of the vocal tract tube. Different sounds are formed by varying the shape of the vocal tract, such as the raising of the tongue. During speech, one's vocal tract is constantly altering its shape so as to produce all the different sounds necessary to speak intelligibly. Thus, the spectrum of the speech signal will vary in time as the vocal tract's shape varies.

The system described below for generating a speech signal can be modeled as a digital system. Figure 9.1 shows a digital model for speech signals. In fact, computer speech

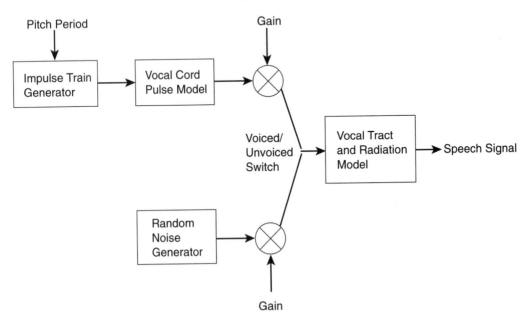

Figure 9.1 Digital Model for Speech Signal Generation.

is often generated by a system like this. This model also sheds light on speech encoding. It is the elements of this system, the pitch, resonant frequencies, etc., that can be digitally encoded. The goal is extract these features from the speech signal to use in a speech encoding algorithm.

9.1 BASIC SPEECH ENCODING

There are two major types of digital speech encoding: predictive coders and frequency domain coders. Predictive coders attempt to model the speech signal at any time instant as a predicted value plus an error term. The most familiar predictive coder is the differential pulse code modulation (DPCM). Figure 9.2 shows a DPCM modulator.

In its simplest form with $B(z) = 0$ and $A(z) = z^{-1}$ (unit delay), the transmitted output $e^q(k)$ is equal to a quantized value of $s(k) - s(k-1)$. The receiver side is shown in Figure 9.3.

A more advanced DPCM than presented above is used for speech encoding and sets up the basis for the CCITT standard at 32 kbps, G.721.[2] The predicted value for this standard can be expressed by:

$$s(k|k - 1) = \sum_{i=1}^{N} a_i \, s(k - i) + \sum_{j=1}^{M} b_j \, e_q(k - j)$$

9.1

In this approach both the quantizer step size and the predictor coefficients are adaptive.

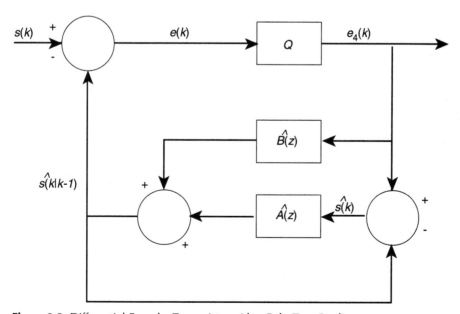

Figure 9.2 Differential Encoder Transmitter with a Pole-Zero Predictor.

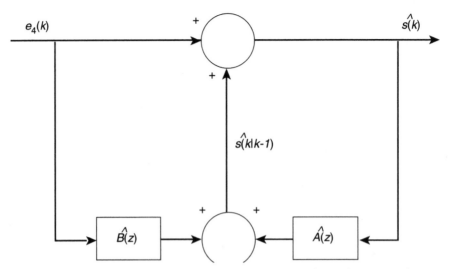

Figure 9.3 Receiver for a Differential Encoder Transmitter with a Pole-Zero Predictor.

However, the above method still leaves us with a bit rate of 32 kbps. This is too high a rate to support an improvement over the analog system. A different type of predictive coder, called analysis-by-synthesis coder, is used to achieve good quality speech at 16 kbps and below. The encoding and decoding steps are shown in Figure 9.4.

A candidate excitation sequence of some block length is applied to a long-term predictor cascaded with a short-term predictor. Typically the long-term predictor is doing pitch determination, while the short-term predictor is usually predicting the spectral envelope or formant. The reconstructed speech sample at time n is subtracted from the corresponding input speech sample and passed through a perceptual weighting filter, and the weighted error is squared. This process is repeated for each sample, and the average weighted error is taken over the block. This whole process is now repeated for all the candidate excitation sequences in a restricted set. The excitation sequence producing the smallest average error is sent to the receiver. If there are K possible excitation sequences of block length L in the set, we must send $(1/L)\ln K$ bits/sample to the receiver.

Analysis-by-synthesis coders are the coders of choice for digital cellular networks because currently they provide the best tradeoff between bandwidth and voice quality. As can be imagined, there are many variations on the scheme described above. These variations will be further discussed when this chapter describes the current and future speech encoding schemes for digital cellular networks.

The other major speech coder is frequency domain coding. There are two basic types of frequency domain coders: transform coders and subband coders. Transform coders take a discrete transform of input speech samples and assign bits to the transform coefficients. These coefficients are then quantized and transmitted to the receiver. At the receiver, the frame of speech is recovered by taking the inverse discrete transform. The most commonly used transform for speech applications is the discrete cosine transform.[2] This transform

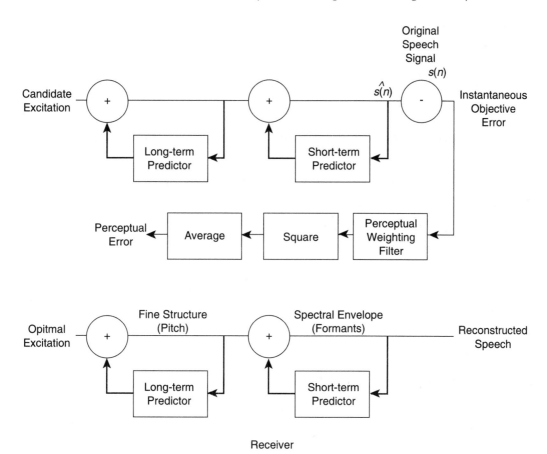

Figure 9.4 Analysis-by-Synthesis Predictive Coder.

allows for perceptually important frequency components to be coded more accurately than others.

Subband coders digitally filter speech into nonoverlapping frequency bands. Each band is then translated down to a baseband and coded separately, using some digital coding scheme, e.g., PCM, DPCM. The receiver simply decodes the bands, translates them back to their original frequency range and sums to reconstruct the speech signal. This method allows for certain bands to be quantized/coded with greater accuracy than other less-speech-important frequency bands.

These frequency domain coders, while conceptually nice, simply do not provide an adequate means for achieving adequate voice quality with bit rates below 16 kbps. For this reason, all the digital cellular networks being used or under consideration use a predictive coding algorithm.

9.2 CURRENT CELLULAR NETWORKS

The analog cellular system used in the U.S. is called AMPS. AMPS uses Frequency Division Multiple Access (FDMA) with channel spacing 30 kHz apart. One full-duplex channel, transmit and receive, will take up 60 kHz of bandwidth. The voice is FM modulated and band limited to under 30 kHz. The total number of duplex voice channels the original system could accommodate was under 400.

With only 400 possible duplex voice channels for the system, an improvement to the system was necessary. A digital network provided for substantial increases in the number of voice channels. Some of the options for a digital network include Time Division Multiple Access (TDMA) and FDMA combinations, time division duplex (TDD) combined with TDMA and FDMA, and Code Division Multiple Access (CDMA). Examples of these are discussed in Chapter 2. TDMA/FDMA combinations are used in the current digital cellular networks around the world: NADC in North America, JDC in Japan and GSM in Europe. Table 9.1 compares these systems. The Digital European Cordless Telephone (DECT) system uses a TDMA/TDD/FDMA approach. Finally, CDMA systems, while currently not in commercial use, hold definite promise for the next generation of digital cellular networks. All of these systems require some form of digital speech encoding. In the next section, we describe the coding used currently in the NADC, JDC and GSM systems.

9.3 CURRENT SPEECH ENCODING TECHNIQUES

The current North American and Japanese systems (NADC and JDC) use a speech encoding scheme called Vector Sum Excited Linear Predictive coder (VSELP). This coder is in the analysis-by-synthesis family of coders and is a modified version of the codebook excited linear prediction (CELP).[3] Figure 9.5 shows a block diagram of a CELP encoder. This technique attempts to mimic the two main parts of the human speech system: the vocal cords and vocal tract. The vocal tract is simulated by a time-varying linear predictive filter (pitch predictor), and the vocal cords sounds used for the filter excitation are simulated with a database (codebook) of possible excitations. In addition, linear prediction computations (LPC analysis) are performed to make the perceptual weighting filter adaptive over time.

The VSELP encoder partitions the speech signal into 20 ms segments. The encoder sequences through the possible codebook excitation patterns and possible filter parameters to find the synthesized speech segment that gives the best match to the original speech segment. The parameters that cause this best match are then digitally transmitted. In the NADC and JDC systems, data rates are only 8 kbps. In the NADC, this allows for a combination TDMA/FDMA system where each frequency band of 30 kHz can now support three time-multiplexed channels. Thus, there is a threefold increase from the analog system. If the data rate could be further reduced to only 4 kbps, six time-multiplexed channels could be supported, thus increasing the channel capacity to six times that of the analog system. New ideas for achieving these low rates are discussed in the next section.

TABLE 9.1 Comparison of Digital Systems

	GSM	NADC	JDC	CDMA
Geography	Europe	North America	Japan	U.S.
Service	1991	1991–1992	1991–1993	1992–1996
Frequency Range	935–960 MHz 890–915 MHz	824–849 MHz 869–894 MHz	810–826MHz 940–956 MHz 1429–1441 MHz 1447–1489 MHz 1453–1465 MHz 1501–1513 MHz	824–849 MHz 869–894 MHz
Data Structure	TDMA	TDMA	TDMA	CDMA
Channel per Frequency	1–16	3–6	3–6	118
Modulation	0.3 GMSK	π/4 DQPSK	π/4 DQPSK	BS/MS QPSK/OQPSK
Speech CODEC	RELP–LTP 13 kbps	VSELP 8 kbps	VSELP 8 kbps	8550 bps
Mobile Output Power	3.7 mW to 20 W	2.2 mW to 6 W		2.2 mW to 6 W
System Spectrum Allocation	50 MHz	50 MHz	110 MHz	50 MHz
Modulation Data Rate	270.833 Kbits	48.6 Kbits	42 Kbits	1.2288 Mbps
Filter	0.3 Gaussian	$\sqrt{raised}\cos ine$	$\sqrt{raised}\cos ine$	
Channel Spacing	200 kHz	30 kHz	25 kHz	1.23 MHz
Number of Channels	124 frequency channels with; 3 timeslots per channel (992)	832 frequency channels with; 3 users per channel (2496)	1600 frequency channels with; 3 users per channel (4800)	10 channels 118 calls/ channel
Estimated # of subscribers year 2000	>20 million			
Source	GSM Standard	IS–54	Spec	Qualcomm

In the other current digital cellular system, GSM, which is the European standard, performs its speech coding by using a different analysis-by-synthesis method. GSM uses an RPE/LTP (Regular Pulse Excitation with Long Term Predictor) coder. This voice coder analyzes speech by splitting it into long-term and short-term blocks. The long-term and short-term blocks are modeled and transmitted as representative codes plus a residual error term. This method is more straightforward than the VSELP method and typically has a

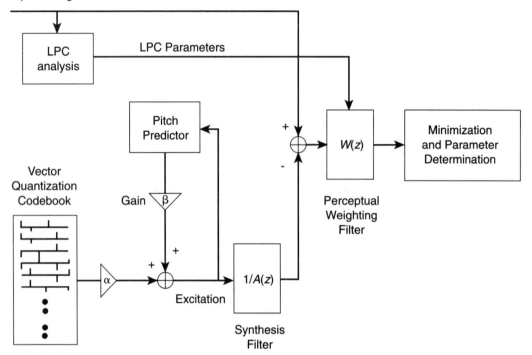

Figure 9.5 CELP-Based Speech Encoder.

shorter delay; however, currently it only achieves a data rate of 13 kbps. As with NADC and JDC, new methods of speech encoding are being developed in order to lower the bit rate.

9.4 FUTURE CELLULAR SPEECH ENCODING

When a cellular system engineer looks at the historical data, it becomes obvious that even the threefold improvements typically gained from the conversion from an analog system to a digital system will not be enough to meet the demand for cellular services in the not so-distant future. Cellular engineers have several avenues such as micro-celling, adaptive antenna arrays and speech encoding techniques, they can take to obtain more channels from the system. Of these, new speech encoding methods hold the most promise for increasing the number of channels on a cellular system. As described above, work is under way in defining standards for speech encoding for the NADC system for a bit rate of 4 kbps. As well, GSM has run competitions of its own for half-rate coders, 6.7 kbps and less.

The challenge is simple: how to send fewer bits while keeping voice quality acceptable and minimizing communication delay. The following paragraphs will discuss some of these new approaches.

One of the main areas of study in speech coding for cellular is in variable rate speech coding. Conventional FDMA and TDMA techniques dedicate a channel or time slot to one unidirectional speech signal, even though a speaker is silent roughly 65 percent of the time. Also, even when speech is present, the short-term rate distortion tradeoff varies quite widely with changing phonetic characters.[4] Thus, the number of bits needed to code a speech frame for a given perceived quality level varies widely with respect to time. Currently, our given digital speech encoders code with a fixed number of bits necessary to achieve quality voice synthesis in the worst-case scenario, even if this case only occurs 25 percent of the time or less. Variable rate coding can achieve a given level of quality at an average bit rate of R_a, which is substantially less than the bit rate of a fixed coder, R_f.

One such variable rate coder maintains an average bit rate of under 4 kbps. This coder, QCELP, is designed by Qualcomm for use in their CDMA digital cellular standard and is a source-controlled, variable-rate coder. QCELP is very similar to the analysis-by-synthesis coders discussed earlier. The difference lies in an adaptive rate decision algorithm. The algorithm keeps a running estimate of the background noise energy and selects a data rate based on the differences between the background energy estimate and the current frame's energy. The possible data rates for this system are 8, 4, 2, and 1 kbps. When the speech signal is in a worst-case situation, the 8 kbps data rate will be selected, just as in the present fixed-rate system. However, when the user stops speaking, the data rate can be reduced. In the case of no speech, a data rate of 1 kbps is used; the 2 and 4 kbps rates are used at other intermediate levels of speech. To accomplish this rate change, the number of updates per block of data of the LPC coefficients, as well as the pitch predictor coefficients and codebook excitation parameters, are reduced or increased as necessary to fit the necessary data rate.

Another type of variable rate coder is called a network-controlled variable rate coder. In this coder, an external control signal switches the data rate to one of a predetermined set of alternative rates. Typically, the control signal is generated remotely in response to network traffic levels. One of the more elegant approaches to this method is called embedded coding. The coder generates a fixed data rate stream; rate switching is accomplished simply by bit dropping. ADPCM is a coder that can make use of embedded coding. A graceful degradation is achieved as the bit rate is dropped. Recently, a method for achieving embedded coding in CELP-type coders was introduced.[5] In fact, the QCELP coder described above included a network-controlled rate feature, where an external control signal can request the 4 kbps data rate for a frame in order to send signaling information.[4]

Implementation of a variable rate coder is another matter that is not at all trivial. Three different multiple access schemes, E-TDMA, PRMA (packet reservation multiple access), and CDMA, all benefit from voice activity detection (VAD) variable rate coding. TDMA and PRMA perform statistical multiplexing of talk spurts. When a particular mobile's talk spurt ends, its time slot is vacated and assigned to another speaker from a mobile in the group with a newly starting talk spurt. In this situation, however, the bit rate is fixed during active speech; there is no graceful degradation as described in the QCELP method. Also, this method requires a lot of overhead signaling to coordinate the rate of switching.

Another interesting method under study for reducing the bit rate in digital cellular networks is called waveform interpolation. As described earlier in CELP coders, the LPC coefficients were computed once per frame, typically 20 ms. The pitch parameters and codebook vectors, however, were updated much more frequently, typically 2–8 ms. The waveform interpolation method relaxes the constraint on the pitch and codebook vector predictors. Through analysis, relatively large deviations from the pitch contour do not affect the perceived speech quality as long as the smoothness of the original contour is maintained.[6] This means that if the update interval of the pitch predictor and codebook excitations is increased, the overall speech quality can be maintained if the values between updates can be interpolated to create smooth functions.

Though this technique is still very much under study, already bit rates as low as 2.5 kbps have been achieved and provide voice quality like that of a vocoder, with very little background noise. While this is still not toll-quality speech, it may have many voice communication applications.

9.5 CONCLUSION

As the cellular systems of today and of the future mature, one constant will always exist; limited bandwidth. Digital speech encoding provides the best way of utilizing this bandwidth to its greatest efficiency. In addition, as new systems such as PCS's and Low Earth orbit satellites (LEO's) are developed, they too will face these same issues and will continue to try to encode speech in as few bits as possible.

REFERENCES

[1] Jayant, N.S. and P. Noll. *Digital Coding of Waveforms.* (Englewood Cliffs, NJ: Prentice-Hall, 1984.)

[2] Hewlett Packard RF Communication Forum, Spokane Division, 1991.

[3] Couch, Leon W., II. Digital and Analog Communication Systems. (Macmillan Publishing Company, New York, 1993), p 464.

[4] Gersho, Allen, and Erdal Paksoy. *Variable Rate Speech Coding for Cellular Networks*, Center for Information Processing Research, Dept of Electrical and Computer Engineering, University of California, Santa Barbara, 1993.

[5] De Iacovo, R.D., and D. Sereno. "Embedded CELP Coding for Variable Bit-Rate Between 6.4 and 9.6 kbits/s," *Proceedings of the IEEE International Conference on Acoustics, Speech, and Signal Processing*, pp 681–683, Toronto, Canada, May 1991.

[6] Kleijn, W. Bastian, and Wolfgang Granzow. "Waveform Interpolation in Speech Coding," Speech Research Department, AT&T Bell Laboratories, 1965.

Chapter 10

Antennas and Power for Mobile PCS

In a PCS, antennas are responsible for the proper transmission and reception of electromagnetic energy that makes communication possible. The pattern, gain, height and location of the mobile unit and cell site antennas are critical factors that must be considered carefully when designing cellular systems. These design parameters are examined to see how engineers have addressed the challenges of achieving effective antenna performance at such high frequencies, with scattered signals and antennas in motion.

Power is another important component in any system. Having a reliable, rechargeable battery pack with a controllable voltage is essential for efficient operation. A brief discussion of power is included in this chapter. The reader should be sensitive to providing reliable power to any PCS.

10.0 ANTENNA DESIGN

Performance and reliability are of paramount importance to customers on a cellular radio system. Confidence in the service quality deteriorates when audible static and dropped calls are frequent. According to Gene Maher of The Antenna Company, a considerable percentage of performance problems are directly related to the design or installation of the cellular antenna. In fact, a large carrier recently ran a cellular radio clinic for about 120 customers who were dropping a high percentage of their calls. The results of the clinic showed that over 100 of the participants had antenna problems that could be linked to the high numbers of dropped calls. [1] Other recent tests performed by cellular carriers have shown that customer complaints of system performance rise dramatically as cellular systems become more saturated. It turns out that the systems were not necessarily poorly designed; the customer's antenna was responsible for the degradation in most cases. [2] As the only communication link between the mobile unit and the rest of the cellular system, the antenna must perform reliably.

10.1 THEORY: POLARIZATION AND GAIN

Antenna polarization describes the orientation or sense of the electric field component of the radiated electromagnetic wave. Vertical polarization is used throughout most cellular systems because it makes antenna mounting easier and provides a good combination of signal range and efficiency. The antennas are oriented vertically, which aligns the electric field with the radiating structure. The radiation patterns for a dipole are shown in Figure 10.1.

The gain of an antenna should always be referenced to another type of antenna (usually an isotropic radiator or a half-wave dipole). Some definitions crucial to this discussion follow:

Isotropic Radiator: The most common reference antenna, it is a theoretical concept consisting of a point source that can radiate equally in all directions.

Directive Gain: In a particular direction, the ratio of the power density (or radiation intensity) radiated in that direction, at a given distance, to the power density that would be radiated at the same distance by a reference antenna radiating the same total power.

Dipole: An antenna fed from the center of the length $\lambda/2$.

Directivity: Maximum directive gain (gain in the maximum-radiation direction).

Power Gain: Compares the radiated power density of the actual antenna and that of the reference antenna on the basis of the same *input power* to both. An isotropic antenna radiates all the power, while some power delivered to the actual antenna may be dissipated.

10.2 MOBILE CELLULAR ANTENNAS

Gain

It is virtually impossible for the mobile vehicular antennas and cell site antennas to always receive direct, line-of-sight waves. As the vehicle moves, the antenna receives both direct and reflected waves from all directions and is therefore required to be almost omnidirectional. To compensate for the low power of a typical cellular phone (3 W), it becomes nec-

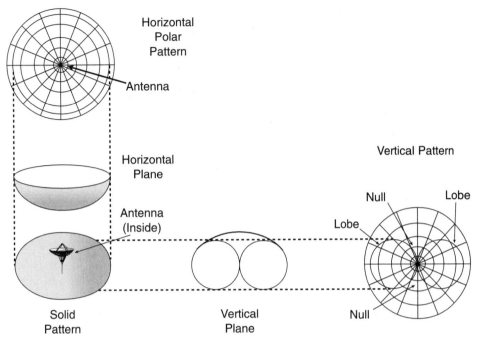

Figure 10.1 Dipole Radiation Pattern
Source: Reprinted from [3].

essary to decrease the vertical component of the radiation pattern and expand the pattern horizontally as shown in Figure 10.2. Power is conserved; it does not increase or decrease—the signal is only being redirected. [4]

Mobile cellular antennas are normally designed to operate as 0-dB, 3-dB, and 5-dB gain antennas (all referenced to a half-wave dipole). The 0 dBd antenna is best suited for use in environments where the cell site towers are located at elevated points or near high

Antenna Gain

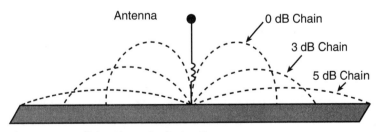

Figure 10.2 Cellular Phone Radiation Pattern

cell site concentrations because the vertical component is not diminished. The cellular system was designed with the 3-dBd antenna in mind because that antenna strikes a balance between effective vertical and horizontal patterns. The radiation pattern is slightly flatter than that of the 0-dBd antenna and provides better performance in weak signal coverage areas. The 5-dBd antenna has an even flatter radiation pattern and is specialized for use in fringe and rural areas where the distance between cell sites is significantly large. It is also recommended for use with .6 W or 1.2 W cellular phones. [1]

At this point it is important to note the reciprocity of antennas; the transmit characteristics are also the receive characteristics. If an antenna is designed to transmit with a severely suppressed vertical component, then it will have difficulty receiving scattered signals with high vertical angles in a mobile radio environment.

Length

Mobile cellular antennas are designed to operate in both the transmit (824–849 MHz) and receive (869–894 MHz) cellular bands and in the future in the 1.8–2 GHz range. Signals are transmitted on one frequency and received on another. These dual frequency channel pairs are separated by 45 MHz. A cellular antenna should perform well over the entire 824–894 MHz range. Increasing the gain of a quarter wave whip antenna (0 dBd) is not only a matter of increasing the height. Increasing the antenna length creates more currents of opposite phase that ultimately cancel the original currents. The result is a taller quarter wave antenna with a new resonance frequency. [5] In order to truly increase the gain, another radiator is added but separated from the original by an inductance (phasing coil) as shown in Figure 10.3. The coil is designed to eliminate the out-of-phase (reverse) cur-

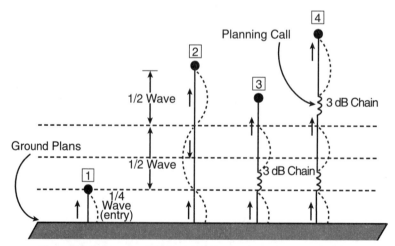

Figure 10.3 Increased Gain Antenna

rent and sum the forward currents to achieve a 3-dBd gain and allows the antenna to be resonant at the design frequency. An antenna with 5-dBd gain must include two phasing coils [5], as shown in Figures 10.3 and 10.4. In the future, antennas will have to operate in the 900 MHz to 2 GHz range.

Placement

Since placement is critical to antenna performance, an overview of the different antenna locations is presented. A key to antenna placement is ensuring that it sits high above the vehicle's roof line to allow it to transmit a clear, unobstructed signal to the cell site tower without interference from the automobile. [6]

Roof Mount:	This antenna performs best because the roof serves as a large ground plane. It is fed directly from the mobile unit but requires drilling a hole in the roof.
Trunk Mount:	The antenna requires an elevated feed to ensure that the mast remains above the roof line for proper performance. It is fed directly from the mobile unit but is not aesthetically appealing.
Glass Mount:	By far the most popular mobile antenna. It provides fair performance and attaches easily to the top of the rear window. No cables or drilling are required because the signal is coupled through the glass. It is not for use in vehicles with metallic-based glass.
Magnetic Roof Mount:	Compact and transportable, this antenna provides good performance but requires a direct feed. It can scratch the paint if improperly removed.
Window Clip:	Installs easily and allows the window to close completely. Highly transportable and is fed from *inside* the car.

10.3 MULTIBAND MOBILE ANTENNAS

Recently, it has become desirable to decrease the number of antennas required for the proliferating varieties of mobile radio communications. Each segment of the radio spectrum requires a different antenna to be compatible with its wavelength. In order to meet this need, some Japanese engineers have designed an AM/FM mobile telephone triband antenna. The structure requires two phasing coils to perform equally well in the AM, FM and cellular radio bands. The top coil, Z_a, is required in the 824–894 MHz band to suppress the current in the top part of the antenna. Coil Z_b allows the antenna to operate as a collinear array with gain by eliminating the out-of-phase (reverse) current component. In the AM/FM broadcast bands, the reactance of Z_a and Z_b is so low that they hardly affect the antenna current distribution. The performance is very close to that of a wire antenna without the phase compensation coils. [7]

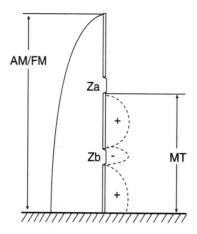

- - - - The current amplitude is the
mobile telephone band

——The current amplitude is the
AM/FM B, C head

+, - The phase relation of the
current distribution

Figure 10.4 Triband Antenna with Coils

10.4 CELL SITE ANTENNAS

Standard high-gain 6 dB and 9 dB omnidirectional antennas are commonly used at the cell site, especially in startup system omnicells. [8] Separate transmit and receive antennas are used and according to Lee [10], each cell can serve 45 voice radio transmitters with three transmitting antennas. Accordingly, two receiving antennas can simultaneously receive all 45 voice signals with each channel employing two diversity receivers. Signals with different fading envelopes are received and then combined, with their correlation determined by the antenna separation. If antennas for space diversity reception are spaced too close together, however, then ripples will form in the antenna receive patterns. [9]

Directional cell site antennas can be used to reduce cochannel and non-cochannnel interference, especially over uneven terrain. A 120 degree or 60 degree corner reflector directional antenna can be used for this purpose. Typical 8-dB directional antenna radiation patterns are shown in Figure 10.5. [10] A top-loaded monopole antenna can be used to create an umbrella pattern with a downward-tilted beam if it is necessary to reduce interference to nearby cells by confining radiated energy to a small area or to provide coverage to a weak spot. [11]

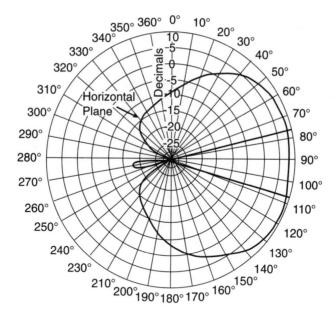

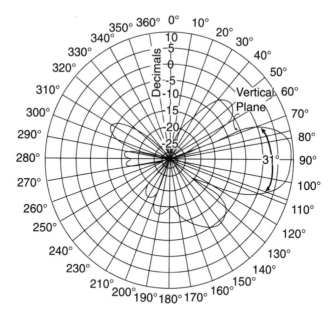

Figure 10.5 A Typical 8-dB Directional Antenna Pattern
(a) Asimuthal pattern of 8-dB directional antenna. (b) Vertical pattern of 8-dB directional antenna.

Source: Reprinted from *Bell Systems Technical Journal*, Vol. 58, January 1979 pp. 224-225.

10.5 POWER FOR PCS SYSTEM

Traditionally, not much is written about power systems. However, it is so important to the operations of a PCS system that some discussion should be offered.

The use of PCS in underdeveloped countries requires batteries that have a long life. In many underdeveloped countries, electrical power is not even available for recharging. Consequently, solar cells or other alternative energy sources are needed to adequately recharged batteries.

This is only one of many difficulties that may arise with PCS power sources. The design of new systems should consider the type of power used, its lifetime and how it can be recharged.

The new Sprint spectrum PCS telephone utilizes a Nickel Metal Hydride (NiMH) battery at 6.9 V. Nickel Cadmium (NiCd) batteries are not used in many of the newer systems. The new batteries have to be charged and discharged regularly to maintain the proper lifetime and operation of the system. Low battery power will cause outages and faulty system operation.

Many sophisticated monitoring and recharging techniques have been established for these newer batteries. In the future, more thought and design consideration should be given to developing PCS power supplies. Longer lifetimes and more accurate batteries are necessary. The uncertainties of power availablity and battery recharge must be given a higher priority in design.

10.6 CONCLUSION

The preceding discussion provided some insight into cellular radio design considerations. Proper antenna design and performance can mean less static and fewer dropped calls, which leads to satisfied customers who have increased confidence in their cellular system. For future PCS, antenna and power considerations in the 900 MHz to 2.5 GHz range will have to be designed. The Sprint spectrum systems has addressed these considerations for the 1.9 GHz range.

REFERENCES

[1] Maher, Gene. "How to Gain Good Reception: What Every Cellular Professional Needs to Know About Antennas," *Cellular Marketing*, November 1993.

[2] The Antenna Company, *Antenna Theory and Application Training Manual*, (The Antenna Company, 1992) p. 2.

[3] Carr, Joseph J. *Practical Antenna Handbook*, TAB Books, 1989), p. 128.

[4] The Antenna Company, *Antenna Theory and Application Training Manual*, (The Antenna Company, 1992), p. 4.

[5] The Antenna Company, *Antenna Theory and Application Training Manual*, (The Antenna Company, 1992), p. 7.

[6] The Antenna Company, *Antenna Theory and Application Training Manual*, (The Antenna Company, 1992), p. 13.

[7] Egashira, Shigeru. Takayuki Tanaka, and Akihide Sakitani. "A Design of AM/FM Mobile Telephone Triband Antenna," *IEEE Transactions on Antennas and Propagation*, Vol. 42, No. 4, April 1994: 42(4):538.

[8] Lee, William C.Y. *Mobile Cellular Telecommunications Systems*, (New York: McGraw-Hill Book Co., 1982), p. 160.

[9] Lee, William C.Y. *Mobile Cellular Telecommunications Systems*, (New York: McGraw-Hill Book Co., 1982), p. 164.

[10] Lee, William C.Y. *Mobile Cellular Telecommunications Systems*, (New York: McGraw-Hill Book Co., 1982), p. 162.

[11] Lee, William C.Y. *Mobile Cellular Telecommunications Systems*, (New York: McGraw-Hill Book Co., 1982), p. 165.

ADDITIONAL REFERENCES

The Antenna Company. *Antenna Theory and Application Training Manual*. (Itasca, IL: The Antenna Company, 1992).

Carr, Joseph J. *Practical Antenna Handbook*. (Blue Ridge Summit, PA: TAB Books, 1989).

Duff, William G. *A Handbook on Mobile Communications*. (Gainesville, VA: Don White Consultants, Inc., 1980).

Shigeru, Egashira; Takayuki, Tanaka, and Akihide, Sakitani. "A Design of AM/FM Mobile Telephone Triband Antenna." *IEEE Transactions on Antennas and Propagation*, April 1994: 42(4).

Jasik, Henry, ed. *Antenna Engineering Handbook*. (New York: McGraw-Hill Book Co., 1961.)

Kraus, John D. *Antennas*. (New York: McGraw-Hill Book Co., 1950).

Lee, William C.Y. *Mobile Communications Engineering*. (New York: McGraw-Hill Book Co., 1982).

Lee, William C.Y. *Mobile Cellular Telecommunications Systems*. (New York: McGraw-Hill Book Co., 1989).

Maher, Gene. "How to Gain Good Reception: What Every Cellular Professional Needs to Know About Antennas." *Cellular Marketing*, November 1993.

Taga, Tokio and Kouichi, Tsunekawa. "Performance Analysis of a Built-In Planar Inverted F Antenna for 800 MHz Band Portable Radio Units." *IEEE Journal on Selected Areas in Communications*, June 1987:SAC-5(5).

Chapter 11

PCS Channel Propagation in Maritime Environments

This chapter presents a detailed overview of mobile cellular communication system performance in maritime related environments. These environmental or "transmission channel" factors need to be better understood because of the growing telecommunication market share of commercial and recreational boaters and fliers. The material presented is based on the large volume of work devoted to this topic by the land-based and shipboard radar development and engineering community in the area of RF propagation near the sea surface. A maritime environment was chosen since a great deal has already been written about other environments (land, air). Since PCS is just beginning to be implemented in the 1.8–2.5 GHz spectrum results of radar systems at these frequencies has been utilized.

An understanding of the propagating medium is critical to the performance of a radar system because of 1) the importance of a radar systems overall mission (the remote detection and tracking of noncooperative targets), 2) the two-way transmission path and associated propagation effects for both monostatic and bistatic system designs, and 3) the significance of the amount of radar energy that is placed on a target in the calculation of received power at the radar antenna. This last point is often summed up as:

"Signal processing, and therefore target detection and tracking by a radar system, first requires a signal to process."[9]

The same statement holds true for PCS communication systems in maritime environments, albeit the loss of signal in a noncommercial communication system is not as critical as it is for a radar or commercial communication system.

The material presented in this chapter augments the empirical test data that describes the transmission channel characteristics in maritime environments. While a 20 dB/decade (i.e., free space) signal roll-off, has been suggested, it has been shown that signal propagation in maritime environments particularly at low antenna heights is highly complex, and a first order estimate of roll-off is closer to a 30 dB/decade value. Note that this value is still less than the 40 dB/decade figure quoted for land-to-land multipath environments. A more detailed analysis of this topic is possible, given the large array of very high fidelity propagation models developed for the study of land- and ship-based military and commercial radar systems.

11.0 THE PROPAGATION CHANNEL

The general block diagram for a communication system [2] is shown in Figure 11.1. As seen in the figure, the "channel" is an integral part of the communication system and can impact the design and overall performance of the system. In Figure 11.1, the general relationship between the information source, transmitter, channel (transmission medium), receiver and information sink (user) is shown.

For PCS cellular telecommunications, the purpose of the transmission channel, is to propagate a bandpass signal (824 to 894 MHz, 1 of 36.4 to 33.6 cm AND 900 MHz–2.5 GHz) between the transmitter and receiver. During propagation, the transmitted signal is attenuated by the channel such that the available signal-to-noise (S/N) ratio at the receiver is decreased. This decrease in available S/N is due to path loss, the presence of channel noise (static in an AM radio) from natural (lightning) and man-made noise (vehicle ignition coil or high voltage transmission lines) sources, as well as from the communication system's receiver (including the antenna). In this context, channel attenuation is not lim-

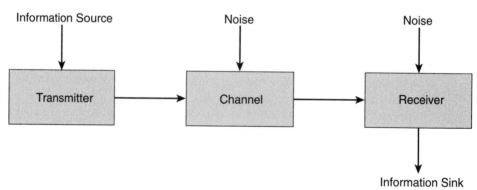

Figure 11.1 General Communications System Diagram.

ited to the traditional definition, which refers to the absorption and scattering of RF energy due to the presence of particulates (clouds, fog and precipitation) in the propagation path.

Channel attenuation can (and often does) vary with time and is a function of transmitter-receiver geometry, the presence of other common frequency emitters and environmental conditions (both local and external). Transmitter-receiver geometry and environmental factors, such as terrain, influence whether a channel can provide multiple paths for the propagation of RF energy between the transmitter and receiver. These multiple pathways arise when the transmitter's antenna simultaneously illuminates both the receiving antenna and another conducting surface (earth, water, buildings, etc.) and each of these pathways has its own amplitude and phase characteristics. These multiple reflections can be approximated and analyzed by simple ray theory.

The superposition of these characteristics at the receiver's location leads to rapid (scintillation) or slow (fading) fluctuations in the amplitude and phase of the transmitted signal.[2,3] Characterizing the signal attenuation by the channel is critical in the design of a communication system because signal attenuation corrupts the transmitted signal and can easily require significant engineering resources for the calculation of "link budgets" and the design of receiver signal processing techniques.

11.1 LINK BUDGETS

Typical link power budget equations show the dependence of the received power (or carrier power (C)) on the transmitter EIRP (effective isotropically radiated power), the receiving antenna gain (G_r), free space path loss (L_p), attenuation loss (L_a) and other system losses. An example of a link equation for a satellite communication system is:

$$C = P_r = EIRP + G_r - L_{ta - Lp - La - Lra} \text{ dBm}$$

11.1

where

P_r = Received power (also called carrier power)

$EIRP$ = $10\log_{10}(P_t G_t)$, transmitted power and antenna gain

G_r = $10\log_{10}(4\pi A_e/\lambda^2)$, effective receive aperture and wavelength

L_{ta} = Transmitting antenna losses

L_p = $20\log_{10}(4\pi D/\lambda$, transmit-receive distance and wavelength

L_a = Atmospheric attenuation

L_{ra} = Receive antenna losses

For a mobile cellular communication system, typical values for the variables in the link budget are shown in Table 11.1 The values for path loss were developed from empirical data and are intended to represent many of the loss terms normally associated with a link budget calculation. This term also represents such dynamic factors as off-axis reception (i.e., the mobile unit is in the antenna's elevation sidelobes), atmospheric refraction effects, surface reflectivity and terrain contour.

Significant theoretical and experimental work has been done to quantify the propagation of RF energy over land and sea surfaces under various environmental conditions. This work has direct application to mobile cellular and microwave datalink communication systems because of the relative heights of the transmitting and receiving antennas and the presence of unwanted multipath reflections, anomalous propagation by variable height and densities of ducts, and signal absorption sources along the propagation path. While this is especially true for the case of a land-based battlefield radar system, the same statement holds true for radar systems operating in the maritime environment.

The radar range equation that is used for the prediction of the received power from a target of a given radar cross section is similar in form to equation 11.1, with the exception of several missing terms used to describe the pattern propagation factor, F, etc., and the two-way propagation path. The purpose of the pattern propagation factor is to account for modifications to the field strength at the receiver's location because of atmospheric refraction, multipath reflections from a reflecting surface within the transmission beam width, and diffraction over this surface.[4] The development of the path loss term and the treatment of multipath reflections is reviewed in detail in Section 11.3.3.

TABLE 11.1 Standard Mobile Telecommunication System Parameters

Parameter	Value
Transmitted Power	Typically 40 dBm (50 dBm maximum)
Transmit (cell site) antenna gain	8 dB (dipole), 120 (az) by 31 (el)
Receive (mobile) antenna gain	0 dB (dipole), omnidirectional
Receive antenna noise	-9 dBm
Receiver noise, B=30 kHz	-120 dBm
Reference minimum detectable signal	-100 dBm
One-way path loss, land	Typically 40 dB/decade
One-way path loss, water	20 dB/decade (i.e., free space) *
Cell site antenna height	Typically 30.5 meters
Mobile antenna height	Typically 3 meters
Range	Maximum of 20 kilometers

*according to Lee's text
Source: Reprinted from[1].

11.2 ATTENUATION

Attenuation loss in any communication system is defined as the difference between the power received under ideal propagation conditions and the power received under nonideal conditions (e.g., precipitation). Ideal propagation conditions are not the same as free-space conditions and take into account the absorption of RF energy because of molecular resonance in the troposphere at altitudes of less than approximately 40 kilometers. This absorption of RF energy due to "clear air" resonance is a function of frequency, with the significant absorption lines being found at 22.235 GHz (water vapor), and between 50 and 66 GHz (oxygen); see Figure 11.2.

Figure 11.3 provides the rates of attenuation at an elevation angle of 0 degrees (i.e., horizontal) at typical communication and radar system frequencies. Note that these atten-

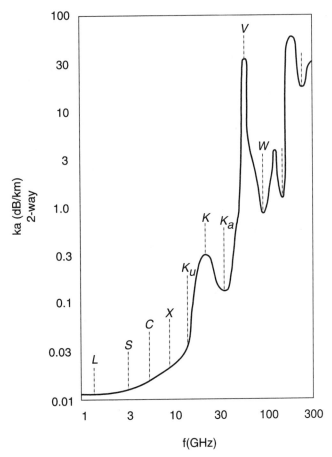

Figure 11.2 Atmospheric Attenuation Coefficient Versus Frequency
Source: Reprinted from [4].

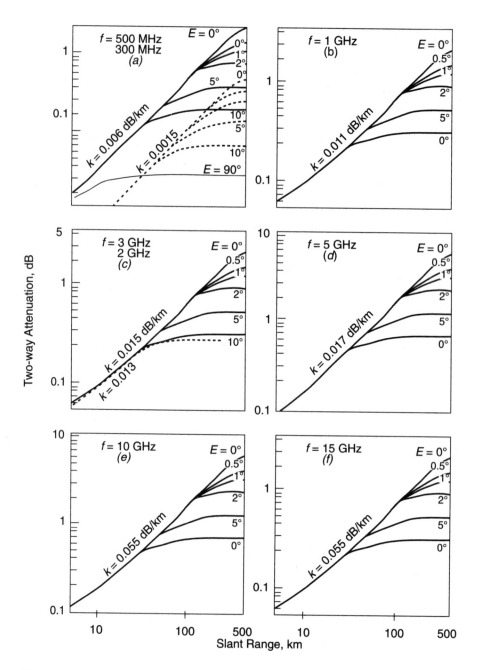

Figure 11.3 Atmospheric Attenuation of RF Energy at Different Elevation Angles
Source: Reprinted from [4].

uation rates are expressed as two-way propagation values. To determine the proper values for a communications system, simply divide the rate by 2. As can be seen in the figure, an increase in the transmission frequency results in a corresponding increase to the "clear-air" attenuation rate.

From the Figure 11.2 it can be seen that the attenuation rate for mobile cellular communications (at approximately 2 GHz) is between 0.01 to 0.02 dB/km. For a maximum transmission range of 20 kilometers, this rate would equal a loss of approximately one-tenth of a dB. This level of loss is not considered to have a significant impact on link budget analysis. The same cannot be said for microwave data links at higher frequencies, particularly in the presence of precipitation.

Communications system links, particularly those above 8 GHz and below 50 GHz (satellite communication systems), must be designed for resistance to attenuation (fading and depolarization) due to rain.[3] Typically, this resistance is achieved by increasing the available signal strength at the receiving antenna and by increasing the EIRP of the transmitter. Whether or not this is pursued is a design tradeoff that must balance such factors as economics (system cost and the costs associated with outages) with the criticality and desired availability of the link.

Attenuation of RF energy by precipitation (because of absorption, scattering, and depolarization effects) is a function of RF frequency and polarization, temperature, and rain rate (drop size and shape). Fortunately for cellular communication systems, the attenuation due to rain is not significant below approximately 2 GHz, and as long as the drops are small compared to wavelength (drop diameter to wavelength ratio), the attenuation is proportional to the 2.3 to 2.8 power of the frequency.[5] In the heaviest rain (greater than 20 mm/hour), drop diameters increase and drop shapes become oblate (flattened at the base) and are no longer symmetrical.

Studies have shown that the limiting diameter in the horizontal plane is approximately 6 mm.[3] The non-spherical shape of rain drops greater than 3 mm in diameter accounts for the depolarization effects observed when orthogonal, linear, polarized RF energy propagates through a rainstorm. In Figure 11.4 the relationship between the one-way attenuation coefficient, rain rate, and frequency, as well as the geographic statistical distribution of various rain rates in the continental United States, can be determined. As can be seen from the figures, the specific attenuation due to precipitation is significant for data links at the higher microwave frequencies (i.e., point-to-point microwave, terrestrial communication and satellite communication systems at 4, 6, and 10 GHz).

RF attenuation by foliage has been studied for some time by the battlefield radar system development community. The objective of these radar system types is to detect personnel and vehicles in the presence of severe clutter source returns (from terrain and man-made structures), scattering, and absorption. Battlefield communication systems, and now commercial cellular telecommunication systems, must also contend with the absorption, scattering, and/or diffraction phenomena associated with foliage in the propagation path. Overall signal loss due to foliage is therefore a complicated topic that is dependent on such parameters as 1) the size of leaves, branches, and trunks, 2) the density and distribution of leaves, branches, and trunks, and the height of the trees relative to the transmitter and

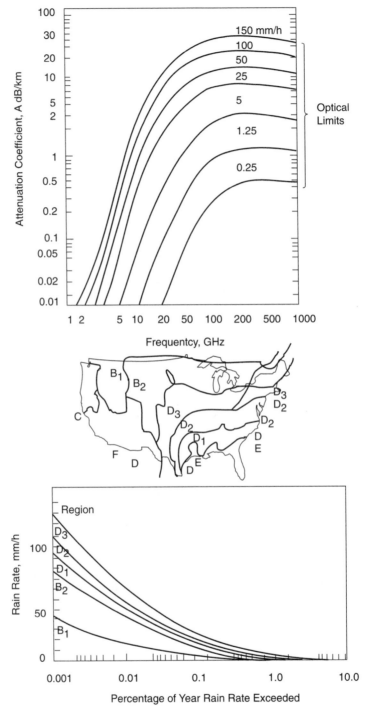

Figure 11.4 Attenuation of RF Energy Due to Precipitation
Source: Reprinted from (3, 5].

receiver antennas, and 3) the overall geometry and depth of the foliage along the propagation path.[1, 6]

Early studies of foliage attenuation at frequencies from 100 to 3000 MHz found that for either orthogonal linear polarization, the one-way attenuation by trees with leaves in that frequency region was given by

$$A = 0.25 f^\gamma \text{ dB/m} \qquad\qquad\qquad 11.2$$
$$[5]$$

where

$$f = \text{carrier frequency in GHz}$$

$$\gamma = 0.75$$

This relationship is plotted in Figure 11.5 for both vertical and horizontal polarizations.

For cellular communications at frequencies of 1 to 2.5 GHz, the foliage attenuation loss is approximately 0.22 dB/meter to 0.35 db/meter, which is significant. There is disagreement over the effect of foliage depth on RF attenuation, but the CCIR has adopted the following relationship for communication calculations where the transmitting and receiving antennas are close to a grove of trees:

$$A = 0.2 f^{0.3} d^{0.6} \text{dB/m} \qquad\qquad\qquad 11.3$$
$$[5]$$

where d is the foliage depth in meters.

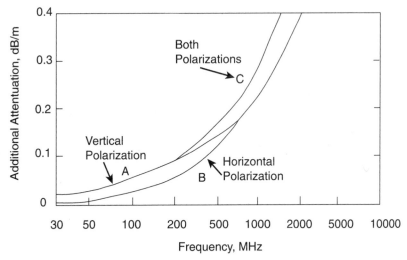

Figure 11.5 Attenuation of RF Energy Due to Foliage
Source: Reprinted from [5].

It is not clear why the frequency exponent should be lower than the value in equation 11.2. Other experiments and tests indicate that RF attenuation due to individual trees or a tree-lined road was 2 dB/meter at 869 MHz and 2.8 dB/meter at 1.5 GHz.[5] These values track with the limited discussion of diffraction and attenuation due to foliage in the class text, which states that there is an overall 20 dB/decade loss over the propagation path when foliage is present.[1]

Cellular communication signals can be attenuated by the presence of objects with dimensions approaching one-half wavelength (dipoles) of the propagating frequency.[1] For mobile communications, this length is approximately 6.9 inches and can easily be approached by small twigs and long pine needles. This source of attenuation can be significant in communications because of the large potential densities of these dipoles. An analogous situation exists for radar systems that are required to operate in environments containing chaff.

Chaff is a passive form of electronic countermeasure (ECM) that dates back to World War II and is generally constructed from aluminum-coated fibers with lengths cut to one-half the wavelength of the frequency of the victim radar. Maximum reflectivity of a chaff dipole (or absorption by a pine needle) occurs when the dipole is oriented perpendicular to the direction of propagation. The maximum reflectivity of a chaff dipole (or attenuation if it is made ideally absorbing) has been found to be

$$\sigma = 0.857\lambda^2, \text{ in square meters} \qquad\qquad 11.4$$

and the collective reflectivity of a group of N randomly oriented dipoles is

$$\sigma = 0.18N\lambda^2, \text{ in square meters} \qquad\qquad 11.5$$

Typical high-density chaff clouds can be represented by a reflectivity of approximately 50 m^2/km^3. From equation 11.4, a single chaff dipole cut for a frequency of 850 MHz has an effective maximum radar cross section of approximately 0.1 m^2, and from equation 11.5, the number of chaff dipoles per cubic kilometer can be found to be on the order of 500. This is a small number of dipoles! The calculation was conducted to explain that the role of chaff in countering a radar system is in the presentation of false targets, not RF attenuation (although there have been reports of the existence of "black" chaff). Considering the density of long-needle pine trees in the mid-Atlantic United States, it is no wonder that attenuation from pine needles has been reported by the cellular communications community.

11.3 LOW ALTITUDE RF PROPAGATION

Most texts discuss two distinct RF propagation "regions" within the troposphere.[6] One of these regions is defined to be the diffraction region, or the region that lies *below* the tangential plane formed by the transmitting antenna and the earth's surface at the geometric horizon. An equation for computing the geometric horizon can be found from simple geometry to be approximately $d = \sqrt{(2kah)}$ where the product ka is the effective earth radius and h is the height of the transmitting antenna. Note that this equation is only an approximation and that h is assumed to be significantly smaller than a.

For the case of standard atmospheric refraction, $k=4/3$; when h is measured in kilometers, this equation reduces to d (in kilometers) = $130\sqrt{h}$. This equation provides the geometric horizon distance when the effect of receiving antenna height is not included (i.e., h_r = 0 meters). With the receiving antenna height expressed in kilometers, the modified geometric horizon range can be found by:

$$d \text{ (in kilometers)} = 130 \; \sqrt{h} + \sqrt{h_r}) \qquad\qquad 11.6$$

Mobile systems beyond this horizon will receive transmitted RF energy by diffraction effects. Note that equation 11.6 includes the condition for standard atmospheric refraction. Figure 11.6 plots this equation for various transmitting antenna heights and a fixed receiving antenna height of approximately 3 meters (from Table 11.1).

Note that in the figure the standard cell site antenna height of approximately 30 meters in a standard atmosphere over flat terrain will prevent the mobile radio from entering the diffraction region. Thus, for cellular communications in a maritime environment, diffraction effects are primarily due to the presence of obstructing objects (islands, large ships, etc.).

However, given a single cell site (transmitting antenna) system operating in a maritime environment, say, from an island, a coverage range of greater than 20 kilometers might be required. In this scenario, if antenna height is limited the potential for mobile radios to be well within the diffraction region, will be the limiting factor on link performance.

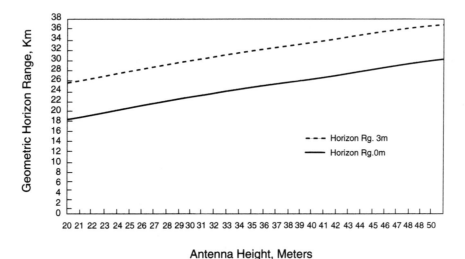

Figure 11.6 Geometric Horizon as a Function of Transmitting Antenna Height

The second propagation region is called the interference region and is *above* the tangent plane formed by the transmit and receive antennas and the geometric horizon. Within this region, the RF signal arriving at the receiving antenna can arrive by multiple "ray paths," which give rise to interference phenomena. The actual received signal power can therefore be predicted by constructive and destructive interference patterns based on the amplitude and phase of the superimposed ray paths. Figure 11.7 provides a graphic representation of these two propagation regions.

Studies on the performance of radar systems against low elevation angle targets in maritime environments have shown the existence of a third propagation region that extends from the geometric horizon point and includes the boundaries of the interference and diffraction regions. Within this region, the RF signal strength at the location of the target (or receiving antenna) is a function of both diffraction and refraction effects and is not easily calculated.[7]

11.3.1 Refraction

Refraction for radar and RF communication systems is defined as the change in the direction of an electromagnetic wave due to spatial changes in the index of refraction of the propagation medium. From physics, the index of refraction, n, is defined to be:

$$n = \frac{c}{v_p} \qquad\qquad 11.7$$

where

c = speed of light (3.0×10^8 meters/second)

v_p = wave phase velocity

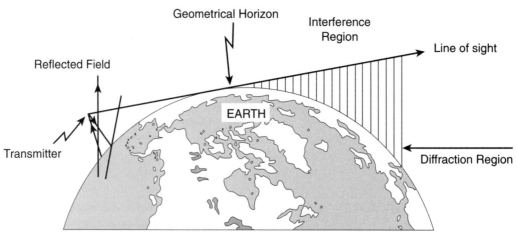

Figure 11.7 RF Propagation Regions
Source: Reprinted from [6].

In the troposphere, the index of refraction normally decreases with increasing altitude. This results in the part of an RF signal phase front (wave) at higher altitude traveling faster than the part in proximity to the surface. This results in a slight downward bending of the RF wave and an increase in the geometric horizon range. This standard atmospheric refraction phenomena is accounted for in geometric ray path calculations by increasing the k (effective earth radius) factor from unity to 1.33.

A standard tool used by radar engineers in plotting the effective elevation beam pattern and rapidly determining the elevation angle of a target at a given altitude is the Blake chart.[7] Charts such as the one shown in Figure 11.8 were developed by the Naval Research Lab (NRL) during early studies on the effect of atmospheric refraction on the coverage of a radar.

Changes in the slope of the refractive index as a function of altitude result in what is referred to as anomalous propagation. Under such conditions RF waves are "bent" in a

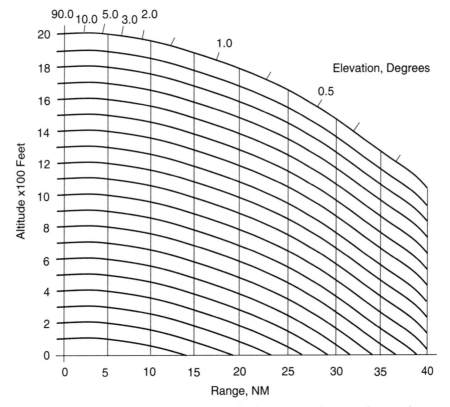

Figure 11.8 Calculated Blake Chart for Standard (Exponential) Atmosphere and Transmitting Antenna Height of 20 Meters

Source: Reprinted from [7].

manner different from that observed with standard refraction. Three general cases of anomalous propagation are possible: subrefraction, superrefraction and trapping (often referred to as ducting).

The treatment of ducting (severe superrefraction or trapping) in PCS is not sufficient to describe the frequency or severity of this propagation condition in the maritime environment or in coastal areas. There are two types of ducts—surface-based and elevated—and as their names imply, they are defined by their height above the surface. Surface-based ducts generally arise from a temperature inversion created when warm air overlies a cool surface. This condition can readily be seen in Denver or Los Angeles as a result of the trapping of smog at the inversion height. Sometimes there can exist multiple inversion layers that are created by "elevated" ducts that are not directly connected to the cool air mass next to the surface.

In maritime environments, another type of duct, the evaporation duct, is created by the humidity gradient (observed as a haze layer) that exists just above the sea surface. The typical evaporation duct height for the coastal mid-Atlantic is approximately 5 to 14 meters. At these heights, the receiving antenna is well within the altitude extent of the duct, whereas the transmitting antenna is above the duct boundary. It must be understood that ducts cannot simply be treated as "waveguides," because they are neither isotropic nor homogeneous. A better conceptualization of an RF duct would therefore be a "leaky" waveguide.

Determining the end result as far as the prediction of received power in the presence of a duct of unpredictable extent in height, azimuth, and range is not an easy task. From the positive perspective, ducting can artificially increase the horizon distance of a given cell site by trapping the RF near the earth's surface. However, even if the duct is present at a given instant, there are no guarantees that the duct will persist for any reasonable amount of time or remain in a favorable position to maintain a communication link. From a negative perspective, the presence of a duct can be detrimental to system performance in that the received power of signals from other cell sites (cochannel interference) will be stronger than would otherwise be predicted.

The effect of different propagation conditions on the coverage volume of a shipboard radar system, and, by analogy, a cellular communication system, is shown in Figure 11.9. In the figure, the effect of antenna height with respect to the height of the duct can be seen. Given the sensitivity to height, it can be postulated that ducting conditions (particularly maritime evaporation ducts) will extend the coverage range of a given mobile radio because its antenna will be directly coupled into the duct. For the cell site antenna, on the other hand, it is more likely that the antenna will routinely be located above the duct height, which will lead to subrefraction, loose coupling of the RF into the duct, and spotty surface coverage.

Figure 11.10 provides statistical data on the average heights and frequency of occurrence of RF evaporation ducts in various portions of the Atlantic Ocean, as well as a worldwide average. Because of complex interactions of air masses along coastal boundaries, these evaporation ducts can move inland and become elevated. This type of situation is routinely found in the Chesapeake Bay and on the lower Potomac, Rappahannock and James Rivers.

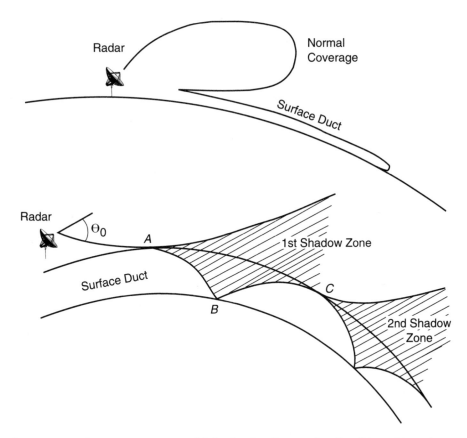

Figure 11.9 Refraction, Trapping, and Subrefraction Propagation Conditions, Normal Coverage, Antenna Within a Surface Duct, and Antenna Above the Surface Duct
Source: Reprinted from [6].

11.3.2 Diffraction

Diffraction is a natural phenomenon that describes a wave's deviation from a straight path (rectilinear motion). For communication and radar systems it is diffraction that allows for the detection of targets below the geometric horizon (Figure 11.7) or behind an object (i.e., in its shadow). By using the Huygens-Fresnel principle that each elementary area on a wavefront can be considered to be an isotropic radiator, the propagation of RF energy in these shadow regions can be predicted. The ability of an RF wave to propagate "around" the geometric horizon or into an object's shadow is a function of frequency. The lower the frequency, the more the wave is diffracted and the higher the signal strength in the shadow (diffraction) region.

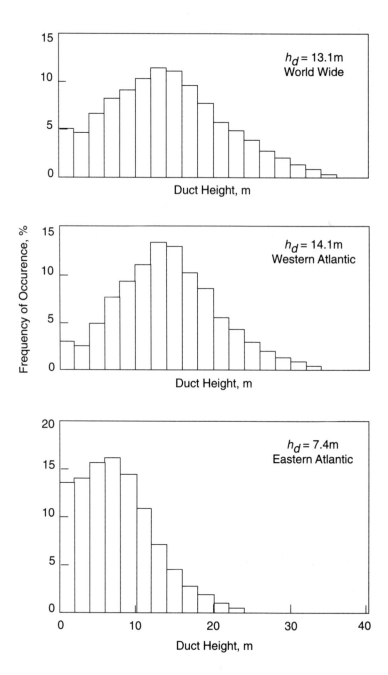

Figure 11.10 Evaporation Duct Height Histograms
Source: Reprinted from [5].

Figure 11.11 is a plot of relative field strength as a function of distance and transmitting frequency for a radar antenna and target altitude of 100 meters. The curves apply to RF propagation over an idealized smooth earth in the absence of an atmosphere (for isolation from refraction or ducting effects).

As seen in the figure, very little RF energy at microwave data link frequencies (higher than cellular communication systems) is diffracted beyond the geometric horizon, because of the relative dimension of the earth as compared to wavelength.

For example, at a frequency of 500 MHz, the one-way propagation loss in the diffraction region is approximately 0.5 dB/km. A communication system operating at this frequency, in hilly terrain, would be required to transmit approximately 10 dB more power than the same system in open, flat terrain (assuming the same desired coverage range of 20 kilometers and everything else being equal). In a maritime environment, given variable

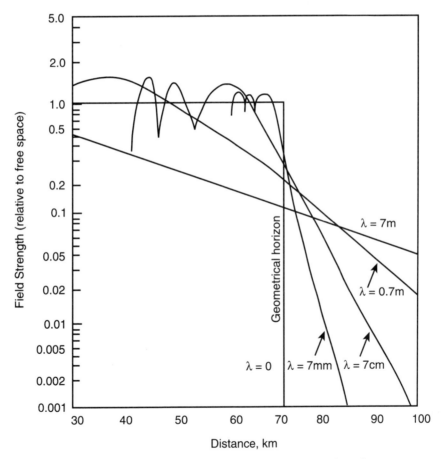

Figure 11.11. Diffraction Theoretical Field Strength Around Earth Surface
Source: Reprinted from [8].

transmit and receive antenna heights, as well as atmospheric propagation effects, a loss of 30 dB/decade appears to be a better initial estimate for link analysis than the often-quoted 20 dB/decade free-space value. The presence of natural (littoral coastline features and islands) and man-made objects (large merchant ships) can also produce diffraction effects that will impact link performance.

Diffraction around objects with smaller dimensions than the earth (such as an island, for maritime systems) is represented by Figure 11.12. In the figure, the Huygens-Fresnel principle that every point along a spherical wavefront of constant phase can be treated as an isotropic point source is represented by point *P*. From physics, the amplitude of the wavefront at any other point due to the contribution of the radiator at *P* can be calculated and varies as $(1 + \cos \theta)$. As shown in the Figure 11.12, the presence of the object also influences the field strength outside the "optical ray trace" shadow region. This effect is why the radar development community has recognized the existence of a third, intermediate, propagation region in addition to the generally accepted interference and diffraction regions (see Section 11.3).

11.3.3 Interference

In the interference region, RF energy can reach the receiving antenna from both a direct ray path from the transmitter and a reflected path, as shown in Figure 11.13. The constructive and destructive interference that arises because there is more than one ray path is called multipath propagation. In communication systems, the effect of multipath propagation is the fluctuation in apparent signal strength of anywhere from no signal to a 6 dB

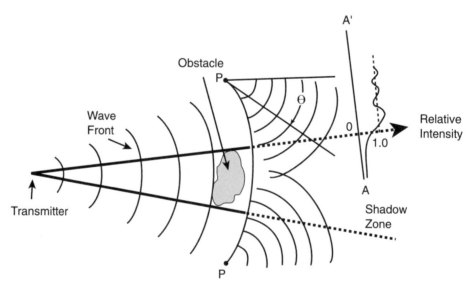

Figure 11.12 Diffraction of a RF Around an Obstacle
Source: Reprinted form [8].

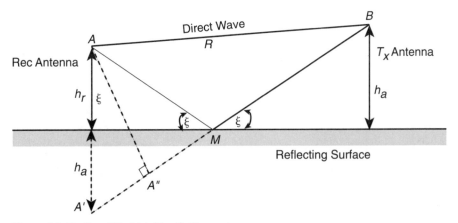

Figure 11.13 Simplified Multipath Geometry
Source: Reprinted from [8].

improvement. In a radar system, the amplitude fluctuation is from 0 dB to 12 dB because of two-way propagation, and the superposition of the direct and indirect paths at the receiving antenna corrupts the elevation angle measurement.

The effective radiation pattern of the transmitting antenna, neglecting atmospheric propagation effects, can be found by using ray theory and geometrical interpretation. Typically, the reflection coefficient is considered to be a complex quantity $\Gamma = \rho e^{j\psi}$, where ρ is real and describes the amplitude of the reflected wave and ψ describes the phase shift of the reflection. At low grazing angles (approaching $0°$) typical of ground-based communication systems, the phase shift of the reflection is $180°$ and Γ is set to -1. Care must be taken in this assumption for vertical polarized systems because the phase decreases significantly as the grazing angle becomes greater than approximately $2°$. In Figure 11.14, the amplitude coefficient, ρ, for various reflecting surfaces, grazing angles, and propagation frequencies is provided.

A geometrical analysis of the simplified multipath propagation case presented in Figure 11.13 reveals that the difference path length between the direct and indirect rays is $\Delta = 2h_r \sin\xi$, when the distance between the transmitting and receiving antennas, R, is significantly greater than the height of the receiving antenna, h_r. For small ξ, the $\sin\xi$ can be replaced with $(h_a + h_r)/R$, so that $\Delta = 2h_r(h_a + h_r)/R$. Since h_a is greater than h_r, this equation can be reduced to $\Delta = 2h_r h_a/R$. The total phase difference between the direct and indirect ray paths, including the phase shift due to the surface reflection, is

$$\psi = \pi + \frac{2\pi}{\lambda} \frac{2h_r h_a}{R}, \text{ in radians} \tag{11.8}$$

The superposition of two signals with the same amplitude (an assumption that is sensitive to grazing angle and surface roughness) will result in a signal whose amplitude varies as $2(1-\cos\psi)$. The power at the receive antenna can then be calculated by

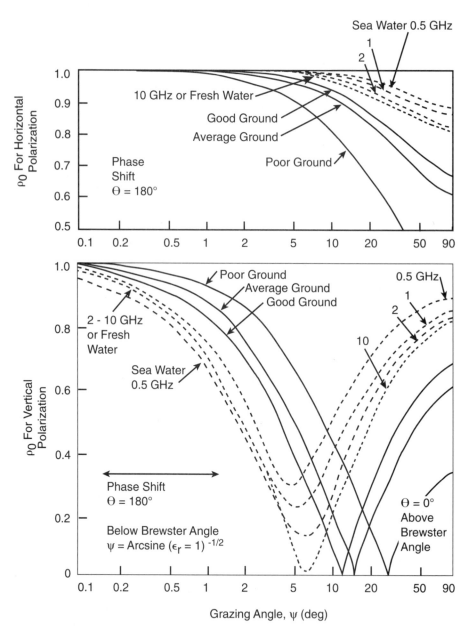

Figure 11.14 Magnitude and Phase of Reflection Over Various Surfaces
Source: Reprinted from [4].

$$Pr = \frac{PtGt}{(4\pi R/\lambda)^2} 2(1 - \cos\frac{4\pi hrha}{\lambda R})$$ 11.9

From this equation, it can be seen that the field strength at the receiving antenna (when it is located in the interference region) will vary from nonexistent (perfect cancellation) to a 4 dB increase (perfect constructive interference) over the free-space condition. By examination, five conditions can be discussed:

1. P_r = zero, when $2(1-\cos\psi) = 0 \Rightarrow \cos\psi = 1 \therefore \psi = 0$

2. P_r = free-space, when $2(1-\cos\psi) = 1 \Rightarrow \cos\psi = \frac{1}{2} \therefore \psi = \frac{\pi}{3}$ and $\psi = \frac{5\pi}{3}$
   i.e., when $\psi = 60°$ or $\psi = 300°$

3. P_r < free-space, when $2(1-\cos\psi) < 1 \Rightarrow \cos\psi > \frac{1}{2} \therefore 0 < \psi < \frac{\pi}{3}$ or $\frac{5\pi}{3} < \psi < 2\pi$
   i.e., when $0° < \psi < 60°$ or $300° < \psi < 360°$

4. P_r > free-space, when $2(1-\cos\psi) > 1 \Rightarrow \cos\psi < \frac{1}{2} \therefore \frac{\pi}{3} < \psi < \frac{5\pi}{3}$
   i.e., when $60° < \psi < 360°$

5. P_r = free-space + 6dB, when $2(1-\cos\psi) = 4 \Rightarrow \cos\psi = -1 \therefore \psi = \pi$
   i.e., $\psi = 180°$

The presence of the reflecting surface therefore causes the effective elevation coverage of an antenna to be broken up into a lobed structure, as shown in Figure 11.15. In the figure, the elevation pattern for a two-dimensional, VHF, horizontally polarized shipboard system is plotted on a Blake chart. From the figure, the effect of multipath interference on the detection capability for a 1 m² target (both pro and con) relative to free-space performance can readily be determined.

Of interest from a cellular communications perspective is the modified elevation coverage for the typical 8 dB transmitting and 0 dB receiving antenna patterns due to multipath lobing and their impact on the viability of the communication link. Utilizing the rule of thumb from radar systems design, the elevation angle of the lowest maximum or lobe, in radians, is $\lambda/4h_a$.

For the case of a mobile radio transmitting antenna height of 3 meters and a λ of 0.35 meters, an elevation angle for the lowest lobe can be found to be 1.7 degrees. Therefore, when the mobile radio is beyond approximately 1 km from the cell site, the cell site antenna will be physically located in the first null (sometimes called the surface null) of the mobile radio. This off-axis transmission loss must be accounted for in the calculation of a roll-off figure for either a land-based or maritime environment. Below we examine the impact of this surface null and the diffraction effects of the geometric horizon on overall link performance in a maritime environment.

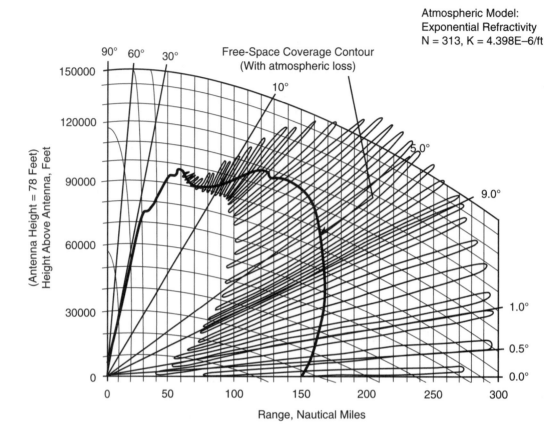

Figure 11.15 Elevation Coverage for a VHF Horizontally Polarized Ship-Board Radar

Propagation prediction within the surface null is very complex, and a fine signal strength lobing structure is observed here as well, particularly at higher microwave frequencies. Figure 11.16 is the calculated vertical lobe pattern in received signal strength over seawater for grazing angles of 4 degrees and 0.7 degrees and a frequency of 35 GHz (Ka Band). The figure demonstrates the relative structure of the RF field at very low heights above the sea surface for significantly higher microwave frequencies. It can be expected that a plot similar to this at ≤1 GHz would not show as much variation in field strength, but the effect of grazing angle will still be observable. As would be expected, a disturbed (rough) sea surface would tend to wash out this structure because of amplitude and phase variations arising from the non-specular nature of the sea surface in this condition.

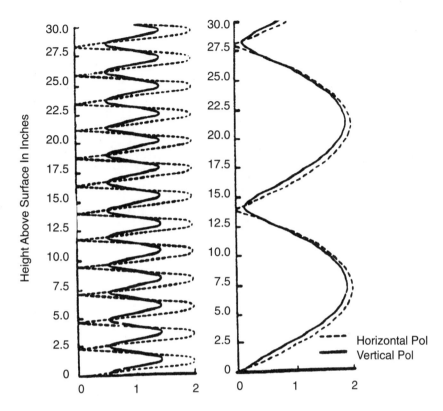

Figure 11.16 Vertical Lobing Pattern at Low Heights Above the Sea Surface

11.4 CONCLUSIONS

Based on experience with radar systems in maritime environments, the general roll-off of the received signal strength at low altitude has proven to be highly unpredictable and unstable temporally. For low altitude targets, the signal strength roll-off of 40 dB/decade has proven to lead to optimistic target-detection range performance. Estimations of the signal strength roll-off in the intermediate region at tactical radar frequencies above 2 GHz have been quoted to be as high as 80 dB/decade. This rule of thumb is based on interpolation techniques between the field strength in the interference and diffraction regions, as shown in Figure 11.17. Again, these values are for the two-way propagation case.

It is therefore expected that for one-way propagation in this environment, a signal roll-off of at least 30 dB/decade, not the 20 dB/decade (free-space roll-off), would be expected.

Since PCS systems in the 1.9 to 2.5 GHz range are in the design stages, conclusions for operations at these frequencies in a maritime environment have been forthcoming from experience with radar systems. Although the conclusions drawn may not prove to be precise, they do offer a good estimate of performance.

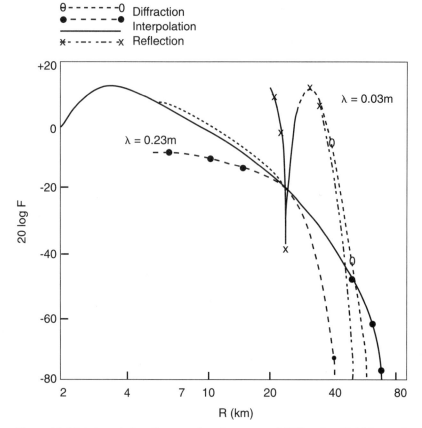

Figure 11.17 Interpolation Between Interference and Diffraction Field Strengths

REFERENCES

[2] L. Couch, *Digital and Analog Communication Systems*, Macmillan, 1993.

[3] T. Pratt and C. Bostian, *Satellite Communications*, Wiley, 1986.

[4] R. Barton, *Modern Radar Systems Analysis*, Artech House, 1988.

[5] W. Lee, *Mobile Cellular Telecommunications Systems*, McGraw-Hill, 1989.

[6] Nathanson, *Radar Design Principles*, McGraw-Hill, 1991.

[7] R. Eaves and R. Reedy, *Principles of Modern Radar*, Van Nostrand, 1987.

[8] C. Blake, *Radar Range Performance-Analysis*, Artech House, 1986.

[9] M. Skolnik, *Introduction to Radar Systems*, McGraw-Hill, 1980.

Chapter 12

Satellite-Based Mobile Communication Systems

Over the years, communication services have been provided primarily through wireline-based networks. The advent of cellular technology over the past two decades has introduced radio systems that provide connectivity between phone lines and mobile users in cars or on foot. However, it is often stated that the ultimate goal of a communication system is its ability to provide pocket-sized, wireless telephones that will acommodate *voice and data* services between any two locations *throughout the globe*. This service would liberate us as communication users from being tied down to a particular fixed location in a telephone network. In addition, intelligent networks would allow for a variety of complex communication technologies transparent to the user. The evolution of this system will be driven by the integration of current *wireline networks* with *future satellite-based* mobile communication systems.

The cellular and personal communications fields have experienced tremendous growth since their inception in the early 1970s. The proliferation of cellular phone systems has been driven by consumer demand for a tetherless communication network. As our society continues to progress toward the information age, the need for cellular communications systems will become even more predominant. Our national highways, especially in highly populated urban areas, are becoming increasingly more congested. Millions of workers across the U.S. spend hours commuting to work everyday. Some of these individuals are currently using existing cellular networks to remain productive during their long commutes. Cellular telephones, laptop computers (designed for car use) and car fax machines allow many commuters to remain productive while sitting idle in traffic. However, these terrestrial networks are already near full capacity and are provided primarily to urban and metropolitan areas. This leaves many rural areas in this country and throughout the world without mobile communications access.

The need for an improved mobile comunication network is unquestionable. The predominant drivers (in terms of performance) for the next generation of personal and cellular communication systems are cost, capacity, universal coverage, spectrum efficiency and portability. The ultimate goal will be to provide ubiquitous person-to-person (as opposed to point-to-point) service that will connect wireline networks with mobile systems while maintaining transparency between users. This goal, once considered only a dream, is on the verge of becoming a reality. Through increasing demand, improved spectrum management and advancements in technology, the concept of a worldwide global communications grid is finally seen as a feasible goal. The evolution of this system will be driven by the need to support global communications as well as the economics of business. Privatization and deregulation will speed this evolution; however, a number of critical issues governing the next generation of PCS need to be addressed today. Comprehensive planning as well as worldwide cooperation are the mitigating factors that must be met if we are ever to see our dreams emerge into reality.

The all-encompassing world of personal communications systems (PCS) includes a broad range of services, including cellular mobile telephones, pagers, cordless telephones and other related wireless technologies. Recently, a number of concepts have been proposed for extending, enhancing and integrating these networks, incorporation of satellites to provide an orbital communications grid, and further utilization of the global fiber optic networks. There are many theories and ambitious ideas on how tomorrow's PCS will operate. This chapter investigates a solution involving the integration of satellites with existing terrestrial networks to meet the demands of our highly mobile society.

12.0 SATELLITE-BASED PERSONAL COMMUNICATION SYSTEMS (PCS)

Current personal communication systems (PCS) were conceived in the early 1970s as a means of increasing capacity, relieving spectrum congestion and reducing blocking rates of single-channel mobile radio systems then in service. Under the cellular concept, allocated channel frequencies can be reused many times over a geographical area. Frequency reuse was provided by dividing a given area into a number of cells. Each cell would contain a terrestrial transmitter that would provide an interface between mobile users and the mobile telephone switching office (MTSO). The MTSO would provide connection to wireline networks. The inventors of this cellular system believed that this concept would have unlimited capacity. They envisioned a system that would divide cells into the subcell system. Over time, problems with cochannel interference, due to the close base station proximity, and analog channel bandwidth caused these systems to fall short of their "unlimited" capacity [1]. Currently, many systems in large urban markets (Los Angeles, New York City, Chicago, Philadelphia) are on the brink of full capacity.

In light of current terrestrial constraints, a number of corporations have proposed the use of satellite-based PCS. Satellites offer a number of advantages over terrestrial-based systems. First, satellites have the inherent ability (due to their high altitudes) to provide coverage to a much larger geographical area in comparison with traditional ground-based systems. In many rural areas in the United States, given the small population associated with these areas and the large areas they occupy, service is made impractical because of the expense of setting up cellular systems. In addition, many developing countries do not even provide the basic telephone service that Americans have grown to expect. A recent survey estimated that 50 percent of the world's population live two hours away from a phone [2]. Both economics and politics prevent many users in foreign countries from accessing basic phone services. Many U.S. firms, like GTE, are using existing satellites in countries with weak or nonexistent infrastructures, e.g., former U.S.S.R., Mexico, and East Germany. The low cost of installation (usually a few medium-sized terminals) make satellite-based PCS very simple and practical solution.

The need for global cellular service in all geographic regions of the world through the use of a single, hand-held headset is a requirement for many business and military users. Future satellite-based mobile communications will meet many of their needs. These orbital constellations will be fully integrated with existing wireline and terrestrial networks. Thus, many incompatibility problems faced by current subscribers (especially in Europe) between cellular systems will be eliminated. Service by these systems will be ubiquitous and will employ a uniform standard that will provide communications to planes, ships, automobiles and to hand-held telephones (only for constellations containing low earth orbit satellites). A single phone located anywhere in the world will be given full communication access. Furthermore, satellite systems will augment existing cellular networks by providing service to roamers (subscribers outside a given service area). Given the nature of satellite systems with their strict timing and position requirements, global positioning to subscribers will be provided. This feature could be integrated in future cars so that preprogrammed maps could provide routes and locations to lost drivers. Low cost of installation, expansion provisions and reliability of service will drive consumer demand for these systems. However, it must be noted that satellite service will be relatively expensive when compared to existing ground-based systems. Thus, these systems are to augment and not replace terrestrial cellular systems. Satellites will offer rural service as well as alleviate congestion in urban areas, but because of their capacity limits, will never replace ground-based systems.

There are two leading constellation design approaches to satellite-based PCS. One approach is to use satellites located in geostationary orbit (GEOs, approximately 36,000 km above the equator); the other approach involves Low Earth Orbiting satellites (LEOs, approximately 500 to 1500 km above the earth's surface). The optimal choice involves prioritizing user requirements. Factors that influence system choice will include cost, extent of service, transmission delays, satellite constellation size, launch expenses, weight, antenna size considerations, receiver sizes, lifetime of system, elevation angles provided to users, capacity requirements, attenuation and cross-polarization factors and spectrum availability. Many of these factors are interrelated; thus, the optimization of a single factor often implies the degradation of another factor. For comparison, two proposed satellite-based mobile communications systems are presented to illustrate the two competing technologies.

The American Mobile Satellite Corporation (AMSC) has proposed a geosynchronous system named MSAT. AMSC along with Telesat Mobile Inc., of Canada are currently designing MSAT to provide PCS to North America. Service will include cellular telephone, flight communication networks, and private service networks. Figure 12.1 delineates the MSAT concept. Coverage will be provided by satellites in geosynchronous orbit. The three-satellite system will utilize directional antennas that will each provide "bent pipe" service through ground-based Ku-band (26.5–40 GHz hubs). Use of Ku-band for satellite-to-hub links allows for optimal allocation of scarce L-band (1–2 GHz) frequencies. Subscribers to this system will place a call through a satellite, which will in turn transpond the received signal to a ground hub. The hub will act as a gateway to wireline networks (assuming the call is mobile user to fixed point user) or will forward the call to a satellite covering the destination user. Figures 12.2 and 12.3 show projected contour gain plots of both the uplink and downlink satellite antennas. Bright regions represent areas of high gain or "hot spots," and darker areas represent regions with little or no gain.

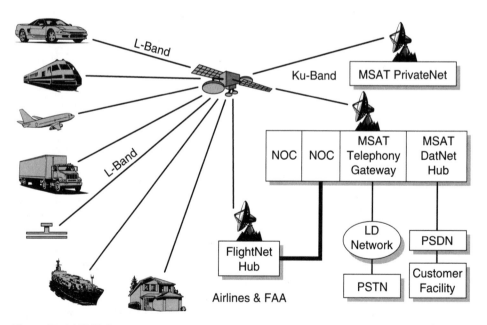

Figure 12.1 MSAT Concept
Source: Used with permission of Dr. Michael Ward, AMSC.

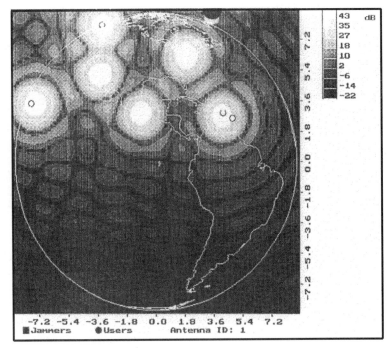

Figure 12.2 Uplink MSAT Spot Beams

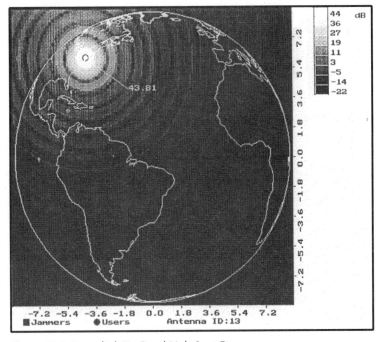

Figure 12.3 Downlink Ku-Band Hub Spot Beam

These plots help illustrate that GEOs, even with their high altitudes, can provide high gain to selective geographic regions. AMSC claims their system will be able to provide EIRP (Effective Isotropic Radiated Power) of 56.6 dBW at the edge of coverage of the continental United States. The model used to generate the contour plots indicates a maximum EIRP of 43 dBW; however, since exact parameters were not provided, conclusive proof will have to be provided by AMSC. MSAT is to provide point-to-point, circuit-switched service, with voice services being offered at 4.8 kbps and data services at 1.2, 2.4, and 4.8 kbps.

The proposed system is not to be a standalone system. Instead, the system will work in conjunction with terrestrial cellular systems. When a subscriber is called from a wireline network, the call will be routed to the MTSO of the subscriber's home cellular system. If the subscriber does not respond, the call is forwarded to a satellite gateway. A subscriber search is then initiated, with the call finally forwarded via satellite to the mobile user. The reverse also holds, assuming the calling party is in an area that is serviced by a ground-based network. Thus, satellites will only service subscribers out of range of terrestrial systems. Hand-offs, transparent to the user, will occur when the user moves in and out of terrestrial networks. This will provide the most effective and efficient use of the allocated satellite resources.

Global coverage by MSAT is not currently planned. However, if successful, the concepts incorporated in MSAT could easily be replicated to provide additional coverage globally. The fundamental problem with MSAT and any GEO-based mobile system is the large distance from the users to the satellites. The size and power requirements of mobile units is a function of the power necessary to communicate with the satellite, the mobile unit's antenna size and the directivity and elevation angle (in reference to the satellite) of the mobile unit's antenna. Most hand-held receivers operate at less than 1 watt. Given large altitudes associated with GEOs and the significant free-space loss experience by the signal arriving from the satellite, the use of pocket-sized portable tranceivers will not be possible because of a lack of power. The power required to communicate with synchronous satellite will be in the neighborhood of tens of watts, which can be generated only by vehicular-mounted radio systems. With the exception of major breakthroughs in battery technology, pocket telephones would have to be remoted to vehicular radio systems.

Low Earth Orbit constellations, or LEOs, involve satellites at altitudes of 500 to 1500 km. As shown in Figure 12.4, LEOs can be placed in polar, equatorial or inclined orbits. An example of a LEO mobile radio system is Iridium. Iridium consists of 66 satellites in six polar orbits, each orbit containing eleven satellites. The satellites will operate at 765 km above the earth and will provide worldwide coverage. Given the shorter transmission paths in comparison to GEOs (thus lower losses due to free space) and the smaller propagation delays, hand-held transceivers will be possible. This implies that truly universal coverage will be provided from any mobile unit to any other phone worldwide. The LEOs will employ phased array antennas to generate 48 spot beams or coverage "cells." Cell-to-cell hand-off between satellites will occur every *nine minutes*. The average number of hand-offs in this system would be greater than that of terrestrial systems, because of the high velocities at which the satellites will be traveling in relation to the earth's rotation. In fact, from a cellular perspective, the mobile user actually acts as the base station

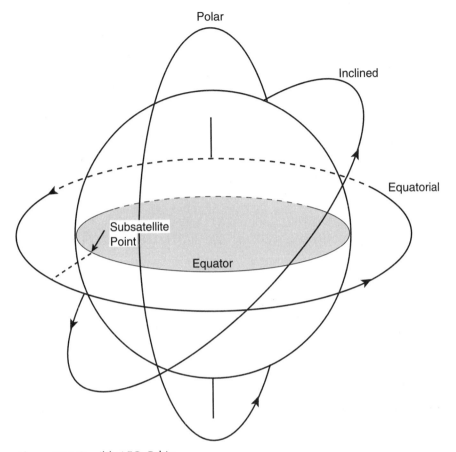

Figure 12.4 Possible LEO Orbits

while the satellite serves as the mobile user[1]. These hand-off algorithms would be complex and would require strict timing and positioning of the LEOs. Given that hand-offs are not the primary focus of the chapter, it is left to the reader to further investigate these algorithms (note Chapter 6). The nonequatorial orbits of LEO satellites provide greater elevation angles to the users. This further reduces attenuation due to small distances traversed between the satellites and mobile users.

Iridium will provide 4.8 kbps voice service with data service offered at 2.4 kbps. Geolocation and paging services will also be provided. LEOs are much less expensive to manufacture and launch. Quick-launch systems already developed can launch a LEO within 72 hours from airplane-based launching systems. This greatly reduces launch costs

[1]In the Iridium constellation, a satellite would be in sight of a given user for approximately 9 minutes, and hand-offs would occur every 6 minutes. If the total time for a call was under 6 minutes, the satellite would act as the base station and the subscriber would be the mobile user.

(from $50 million for GEOs to $9 million for LEOs). This, however, is balanced by the larger constellation needed to provide equivalent geographic coverage.

Even with all their attributes, LEOs do not offer the perfect panacea of mobile radio sytems. Besides the larger constellation (thus, added costs) of LEO systems, there has been a lack of available L-band radio frequencies within which these systems operate. However, recent World Administrative Radio Conference (WARC) has alleviated this problem. Frequency allocations in VHF below 1 GHz were recently released, which allows for the progression of these systems. Once in place, these systems will provide rural service and augment cellular systems in urban areas. Thus, these systems will provide universal services only to subscribers that require these type of services. The associated costs to individuals using LEO PCS will ensure that terrestrial systems maintain their viability while ensuring the efficient use of satellite resources.

12.1 MODULATION SCHEMES FOR FUTURE SATELLITE-BASED PCS

As mobile communication systems continue to emerge, PCS will undoubtedly transition from analog-to digital-based systems and satellite systems must be compatible with these systems. Digital systems will allow for universal compatibility in areas where incompatibility currently exists because of new digital standards. Although there currently exists a uniform standard for North America (Advanced Mobile Phone System AMPS), this is not the case in European countries. Multiple cellular systems exist in Europe. Each national border usually implies a new cellular system and, in turn, a new standard. These multiple standards render telephones useless in adjacent systems. For this reason, Europeans are currently in the process of moving to a digital-based system, which will provide compatibility across borders.

Tables 12.1 and 12.2 show the different analog and digital standards being implemented or being proposed for future systems. It is important to note the power requirements of portable phones and the increased channel capacity of digital systems. These will be the driving factors for the next generation PCS.

As digital systems continue to be planned, the layman may ask, Why digital? Given the success of commercial FM systems, why move to digital based systems. The answer is a bit complex. To begin with, current analog PCS are near full capacity, and increasing the capac-

Table 12.1 Analog Standards

PARAMETER	AMPS	MCS	NMT	C450	TACS
Freq MHz					
Forward	869–894	870–885	935–960	461–466	935–960
Reverse	824–849	925–940	890–915	451–456	890–915
Multiple Access	FDMA	FDMA	FDMA	FDMA	FDMA
Channel BW, kHz	30.0	25.0	12.5	20.0	25.0
Total Channels	832	600	1999	222	1000

Table 12.2 Digital Standards

SERVICE	GSM 1991	NADC 1991–1992	JDC 1991–1993	N-CDMA 1992–1994	B-CDMA 1994
Frequency MHz	935–960 890–915	824–849 869–894	810–826 940–956 1429–1441 1447–1489 1453–1465 1501–1513	824–849 869–894	824–849 864–894
System	TDMA	TDMA	TDMA	CDMA	CDMA
Ch./Freq.	8-16	3-15	3-6	118	500
Modulation	0.3 GSMK	π/4 DQPSK	π/4 DQPSK	BS/MS QPSK/OQPSK	Partially DQPSK
Speech CODEC	RELP—LTP 13 kbps	VSELP 8 kbps	VSELP 8 kbps	8550 bps	All
Mobil Power Out	3.7 mW to 20 W	2.2 mW to 6 W		2.2 mW to 6 W	2.2 mW to 3 W
Allocation	50 MHz	50 MHz	110 MHz	50 MHz	50 MHz
Mod. Rate	270.833 kbps	48.6 kbps	42 kbps	1.2288 kbps	8 Mcps
Ch. Spcg.	200 kHz	30 kHz	25 kHz	1.23 MHz	None
Number of Channels (Initial)	124 freq x 8 per ch = 1000	832 freq x 3 per ch = 2496	1600 freq x 3 per ch = 4800	10 freq x 3 per ch = 1180	500 +
Standard	GSM	IS–54	RCR Spec	None*	None**

ity of these systems center around two alternatives: decreasing channel bandwidth or decreasing cell sizes. The latter would require additional base stations for new cells, which implies additional costs. In addition, as cell sizes become smaller and base stations move closer, cochannel interference becomes a very dominant factor. Changing channel bandwidths would require modification of existing base stations, which would also prove to be very costly and impractical. Digital systems, through coding and encryption techniques, better utilize channel capacity. Digital devices such as computers, faxes and future ISDN wireline networks will interface better with digital PCS. Furthermore, digital systems lend themselves to very effective encryption by virtue of their modulation schemes and multiplexing. One of the major complaints of current analog systems is their lack of privacy and vulnerability to interference. Relatively inexpensive scanners can be illegally used to monitor cellular conversations. Analog encryption, although possible, is difficult and not cost effective to implement.

A number of digital standards have emerged in response to the strong demand for digital systems. Table 12.2 outlines these standards. As mentioned earlier, the European community is quickly moving toward digital systems because a the large number of different (incompatible) analog systems currently exists. The standard adopted by the Europeans, called Groupe Special Mobile (GSM), will be independent of existing analog systems. Subscribers will have to acquire new equipment that will not be compatible with older analog systems. It is hoped that analog systems will be completely phased out of the European community by early next century.

In the U.S., the future of digital systems will be fully backward compatible. This decision by the Federal Communications Commision (FCC) provides for digital systems to operate at the same frequency allocation as existing analog systems. The standard adopted by the U.S. mobile community is the North American Dual-Mode Cellular System (NADCS). The NADC Time Division Multiple Access (TDMA) standard will operate in conjunction with AMPS. These dual modes will contain both analog and digital base stations. For example, when a subscriber initiates a call, the portable transceiver will identify itself as a dual-mode subscriber to the base station. It will then be the responsibility of the MTSO to assign the call to a digital or analog base station. Calls could be changed from analog to digital and back to analog as they progress through various base stations. These future systems will provide regular users with clear, private, dependable digital service, while allowing the casual user access to less-expensive analog systems.

Both the GSM and NADC standards are based upon TDMA/FDMA modulation schemes. This standard will be compatible with future satellite systems. Digital satellite TDMA systems digitize voice and send data bursts into predefined time slots. This allows multiple, time-division conversations to occur over a single FDMA channel. Channel capacity will be determined by voice coder/decoders that digitize analog voice. These encoders include error correction, equalization and interleaving that allow for efficient use of channel.

A number of opponents of the FDMA/TDMA standards have proposed a Code Division Multiple Access (CDMA) alternative. CDMA is a modulation scheme based upon spread spectrum techniques. There are two basic forms of CDMA: direct sequence (DS) and frequency hopping (FH). Direct sequence involves the spreading of a carrier over a much wider frequency band by mixing a high data rate, pseudonoise sequence with the desired signal. The signal appears as noise to other receivers since it's spread below the noise floor. This in turn permits the encryption and privacy that are not currently available in analog systems. Frequency hopping hops the carrier frequency throughout the wideband channel; hence, unauthorized individuals tuned to a specific frequency are unable to listen to the private conversation.

CDMA employs either DS or FH techniques. The code division name comes from the fact that the intended receiver must know the spreading code a priori. Orthogonal sets of codes allow the sharing of bandwidth without interference problems. Satellite digital systems will employ these techniques to spread multiple users over wideband channels. Capacity increases of 10- to 20-fold with respect to AMPS capacity are predicted.

Although no adopted CDMA standard exists, a number of corporations are seriously investigating these modulation schemes. One major concern seen with CDMA is the near-far problem. This occurs when a transmitter close to a DS receiver produces significantly more power at the receiver than the intended transmitter and interference problems areas. Thus, it is important that tight RF power management is maintained in systems employing DS-CDMA.

12.2 DISCUSSIONS AND RECOMMENDATIONS

Satellite-based mobile systems offer a very comprehensive and viable solution to the highly mobile communications needs our society demands. They are not, however, without their disadvantages. Their capacity limitations, both RF power and bandwidth, will not allow for a standalone, satellite-based system. They are meant to augment existing terrestrial systems through the alleviation of congestion in urban areas and the offering of service to rural areas. Cost of service will be expensive (relative to existing systems), given the high cost of launch, constellation size and engineering and maintenance costs associated with satcom networks. Complex hand-off algorithms will have to be implemented, and self-interference problems between closely spaced satellites must be addressed. Limited bandwidth will not economically permit high BW transmissions, e.g., imagery, interactive video, multimedia.

Relative motion between the satellite and the mobile user will cause spectral changes (Doppler effect) in the transmitted waveform that can cause additional degradation to the link. A major concern being investigated is the effect of time-varying mutlipaths that will appear in the transmission of electromagnetic fields from fixed points to mobile terminals (i.e., vehicles). The multipaths will act to provide fading and frequency shifting and will lower overall link performance. The multipath problem is illustrated in Figure 12.5. There are three basic components of the received signal: the direct component, which is the line-of-sight modulated field; specular component, the field received from a single reflection that is a delayed version of the direct component; and the diffuse component, the combined effect of multiple reflections. The specular component is a phase-shifted replica of the direct component and is usually received at a lower level due to the reflection loss during

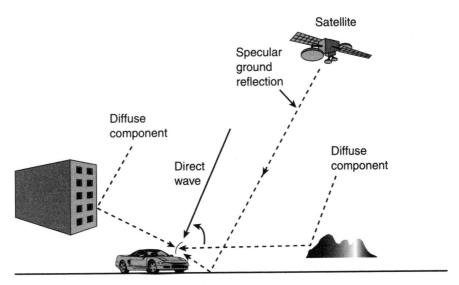

Figure 12.5 Multipath for Satellite Links

propagation. The strongest specular components are generally due to nearby ground reflections [3] and are generally received outside the upward-looking gain pattern of the vehicle antenna. Thus, as a general rule of thumb, specular components are ignored in multipath analysis. The diffused component is composed of all field reflections from all directions (see Figure 12.6). The total of these reflections combine to produce a non-negligible interference to the direct signal. A parameter called the Rice parameter, r, is defined as [Power in the direct component/Power in the diffuse component]. As r approaches zero, the channel is referred to a Rayleigh faded channel. As r increases, the channel is transformed into

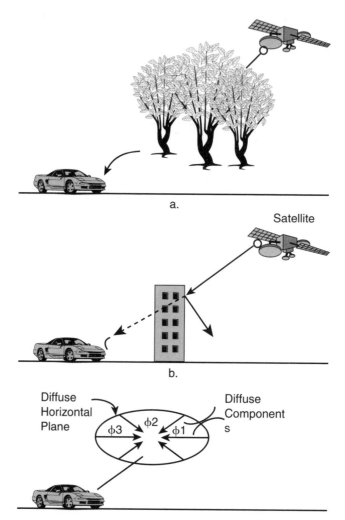

Figure 12.6 Shadowing and Diffuse Component Problems in Satellite Links

a Rice channel. The Rice parameter is a function of reflective terrain in the vicinity of the vehicle antenna and is highly dependent on the elevation angle to the satellite. Thus, the choice of antenna provided on the vehicle is important if Raleigh fading is to be negated. Figure 12.7 presents a number of possible antenna configurations. Also, urban areas with large amounts of reflectors tend to Raleigh fade cellular channels. The Rice factor is highly dependent on elevation angle to the satellite. Raleigh fade margins must be included in any satellite to mobile user design if accurate results are to be obtained.

Shadowing is the blocking of the direct wave due to obstacles (see Figure 12.6). Shadowing generally does not present major problems, since outages are generally temporally nonstationary and occur only intermittently during the vehicles motion [4]. Shadowing can degrade the link by an additional 3 to 10 dB, but only for a short period of time.

Despite their faults, satellite-based PCS have a number of key features. They provide truly universal coverage from any point to any point. Their high cost will be negated through revenue from roaming subscribers and service to developing countries. In addition, thanks to their encryption, many military requirements will be met by these projected systems. Some of these systems will have capabilities to provide digital radio services, which will further increase revenue. Once established, service to new and remote areas can easily be accommodated, and the ultimate goal of providing ubiquitous communication access

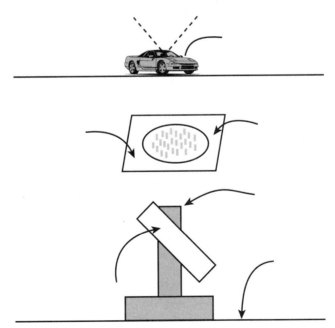

Figure 12.7 Various Antenna Configurations for Vehicular Communication with Satellites

through pocket-sized phones will be met. As our highly mobile society continues to grow, these systems will prove their ability to penetrate the wireless communications frontier.

12.3 LOW EARTH ORBIT (LEO) SATELLITE BASED SYSTEMS

"More than half the world's population lives more than two hours from a telephone," according to Teledesic documents. "Nearly 58,000 villages in Indonesia, 535,000 villages in India, and 151,000 villages in Africa have no telephones." [3] This is one of many reasons there is a race to implement Low Earth Orbit (LEO) satellite communication systems. Besides providing service to remote areas in developing countries, the systems will provide travelers with a means to communicate without the inconvenience of poor-quality phone service in underdeveloped nations. It will also enable international business travelers to be reached anywhere in the world via one telephone number.

As stated in the *New York Times*, "Efforts to forge worldwide telecommunications links, either by satellite or by fiber optic technology, have intensified as governments and private companies seek to develop high-capacity networks that can deliver various forms of data, text. audio, and video into homes and business through a single line." Many LEO satellite systems for personal communications are in various stages of design and production. These systems, using low to medium altitude orbits, will provide the ability to communicate via low power handsets all over the world from the jungles of South America to the deserts of Africa to the mountains of Asia.

To implement these systems, worldwide standards and cooperation are needed. The first step toward regulation was taken in March of 1992 by the 127-nation participants in the World Administrative Radio Conference (WARC-92), which produced an agreement to provide radio frequency spectrum for LEO satellite systems. Systems providing low-cost data communications using small, less powerful terminals and satellites are considered to be "little LEOs" and will operate in the high frequency (VHF) band of 148–150.05 MHz. The "big LEOs," which will provide voice capabilities, were authorized an uplink band of 1610–1626.5 MHz and a downlink band of 2483.5–2520 MHz. In addition, WARC-92 stated that the 1.6–GHz uplink band is also used by the Global Navigation Satellite Systems, both the U.S. Global Positioning Systems and the Commonwealth of Independent States Glonass system, which the new LEO systems must not interfere with. A resolution was also adopted to have the International Telecommunications Union develop standards governing the operation of LEOs, stressing that only a limited number of such systems providing global services can coexist in a given frequency band. Standards are needed to coordinate and share such systems.

The concept for LEO systems was pioneered by Motorola in the late 1980s. Since then, several other systems have been proposed for both worldwide and regional access. The monetary payoffs for working LEO communication systems will be enormous. The local cellular networks are still profitable for those involved. For example, according to the May 10, 1993, issue of Forbes magazine [20], Hillary Clinton had invested $2014 in 1983 into a partnership applying for cellular telephone license. When it was sold five years later to McCaw Cellular, her profit was about $46,000, but if it was sold in 1993, her profit would have been $500,000. FCC Chief of Staff, Blair Levin, has said, "While we can't comment on particular filings, the FCC is continually receiving filings that demonstrate that telecommunications can enhance economic growth at home and abroad." According to the *New York Times*, all of the projects "imply a bullishness about space communications unlike anything seen since the early 1960's, when satellites made international phone calls and television broadcasts part of everyday life. The new systems are appealing: huge parts of the world, from villages in India and China to much of Africa and Latin America, have no basic telephone service and certainly have no advanced forms of communications. Wiring all the world's far-flung outposts would cost billions of dollars, but satellite beams can sweep over deserts and mountains as easily as over urban business centers."

While the technical hurdles are not insurmountable, challenges do exist. The more difficult issues will probably come from the international arena where politics, diplomacy, and economics will be involved.

12.3.1 Challenges - Nontechnical

Since the satellites are in low earth orbits and are not geosynchronous, they are by nature global systems. This means that to be successful, the cooperation of foreign governments is needed. According to an editorial in the *Economist*, [14] global phone systems will need government support to assist in building a large base of subscribers, collecting their bills, issuing licenses to sell services in each country, and setting rules to delineate the terms on which they can compete with local monopoly operators. Foreign governments may be opposed to LEO systems because they would take business away from the state-owned operators, which are a useful source of hard currency in developing countries. Telephone companies usually charge high fees to foreign operators for receiving incoming calls. Also, according to the *Economist*, "engineers with exclusive powers to dispense new lines can fill their own pockets, too: when Telefonica Argentina was privatized in 1991, thousands of unregistered black-market lines were discovered." Most services are attempting to win over reluctant governments by offering to give them a share of the revenue on calls leaving their territories and the rights to market the handsets and collect the bills themselves. "The most skeptical governments will be those that like bullying their citizens. For them, control over the flow of information is a vital way to control society as a whole. In the old Soviet Union, a closed 'key' system linked favored apparatchiks. Only owners of keys to unlock sets could place calls; directories were a prized rarity. Once publicly available, however, telephones give dissidents the power to organize, persuade and publicize."

The WARC-92 helped to eliminate a main political hurdle by setting a worldwide frequency allocation for these systems. Each country is still responsible for granting licenses for the right to operate in their territories. Some systems have international telecommunication partners that will assist in obtaining the necessary licenses overseas, and others are counting on the shared revenue option to open doors for them. In the U.S., the Federal Communications Commission is in the process of granting licenses. From the *Wall Street Journal*, [13] "Some Iridium backers are among those who believe the FCC will ultimately issue no more than two licenses for such systems, but Loral chairman, Bernard Schwartz, said he is closer to that process 'than most people,' and at least three such systems are likely to win government approval. Mr. Schwartz, who has supported the president politically, has close ties to the Clinton administration."

Time is also presenting a problem for the LEO systems. A recent report by the Department of Commerce shows that most countries in Asia, Africa, Latin America and Eastern Europe are in the process of implementing cellular systems to provide basic phone service as well as modernizing existing wired telephone networks. The LEO systems will have to be operational and less costly than these improved local services to compete with them. In addition, the FCC's "Notice of Proposed Rulemaking" states that those granted a license must begin construction of the first two satellites within one year, complete construction within four years and have the entire system in place in six years. The rules of the marketplace will also come into play. The systems that are operational first will be able to gain the largest share of the market.

As the systems begin selling their services, they will want to be able to assure the buyers that their conversations or data transmission will be secure and others will not be able to listen in. If an encryption system is too weak, it will not be effective against other listeners. If it is too strong, governments will oppose it, since they will want the right to listen in. As discussed in the January, 1993, issue of *New Scientist*, [16] the new European all-digital cellular phone system, which was developed by Groupe Special Mobile (GSM), may be blocked from export by Britain's Department of Trade and Industry (DTI). Groupe Special Mobile, a consortium of European manufacturers and telecommunication authorities, developed the system and standards in the 1980s to be used in Europe, which would allow travelers "to use the same portable phone anywhere in Europe and be billed back home." GSM built in an encryption system, called A5, with the assistance of British Telecom, which enabled the British government to have rights to control its use. The FBI in the U.S., as well as the British government, are alarmed by this development because the system is so good that they may not be able to listen in to criminals who are using mobile phones. As a result, the DTI has asked that the GSM standard be changed to eliminate or water down the encryption system. Manufacturers must redesign chips which will take time. Also, it will probably result in two different GSM standards, which will limit the amount that travelers can use their mobile phones internationally.

12.3.2 Challenges—Technical

None of the technical challenges facing LEO systems are insurmountable, although they are too numerous to mention in one report. Because satellites move relative to an earth-based location, a whole set of technical problems not experienced with land-based systems arises. One such problem would be user location. If the user is moving and the satellite is moving, how does the system pinpoint the location and route calls to the position? According to *Aviation Week and Space Technology*, radio determination satellite service (RDSS) is one method to determine the location of the user. It operates by measuring the transit times for a user signal to be relayed to an earth station via two or more different satellites. Iridium plans to determine user position by a small Global Positioning System receiver in each handset. In addition, how will the user be billed for that call; will it be based on the user's location, the home location of the user or some other factor? RDSS may be used to determine the nearest Earth gateway terminal in order to share revenue with the appropriate gateway terminal owner.

Another problem that is complicated by moving satellites is hand-off from one satellite to another. Not only can the user move out of a satellite's coverage area or "footprint," but the satellite may move out of the user's range. Some systems will have communication between satellites play a vital role in the hand-off process. Other systems plan to have large satellite "footprints" to limit the number of hand-offs.

Currently, neither WARC nor the FCC has implemented standards regulating transmissions. The three choices will probably be TDMA, CDMA or FDMA. Each of the choices is being represented by one or more of the proposed projects. For example, Globalstar, Odyssey, and Ellipso will be implementing CDMA. Iridium and Aries plan to use TDMA. Whatever technique is decided upon, the other systems may fight it in court. It is a difficult issue that may never be resolved.

Construction and deployment of these systems present a challenge that will be easier for some systems than for others. For example, the Teledesic system plans to put 840 satellites in orbit in less than two years, something that has never been done before in such a large volume. According to *Scientific American*, [29] "The estimated number of components required for the total complement of Teledesic satellites would be sufficient to create an industry: the startup's application to the FCC cites the need to make 500 million gallium arsenide chips, more than have ever been manufactured commercially." Also, managing, troubleshooting, maintaining, and fixing satellites is more difficult than repairing a land-based cell site. As Jonathan Catherwood of Booz, Allen and Hamilton said, "It's one thing when there are problems in a cellular tower. It's another when there is a problem 460 miles overhead."

12.4 LEO SYSTEMS

Included below is a brief description of some of the emerging LEO systems.

12.4.1 Teledesic

Teledesic was described in *Scientific American* [29] as "an appeal to the utopian strain in the national character and a confidence that no vision is beyond the reach of innovation and hard work. But like many such utopian visions, it also manifests a grandiosity that pushes sanity—in this case, the engineering and fiscal varieties—to its limit."

This system is backed by William H. Gates of Microsoft Corporation and Craig McCaw of McCaw Cellular Communications, Inc., who each own 30 percent in the venture, although the idea was put forth by a venture capitalist, Edward F. Tuck. They plan to put 840 satellites into service by 2001 at a cost of $9 billion. About offering telecommunication services, the *Washington Post* [30] states, "Teledesic does not plan to sell directly to individual customers. Instead it is seeking partnerships with local or national telephone companies here and abroad that would lease space on the Teledesic network to provide phone service in a region." Mass production of the satellites is expected to hold down costs, and technology developed under the "Brilliant Pebbles" defense program should make it easier to manufacture and manage 840 satellites.

As described in *Aviation Week and Space Technology*, "The proposed constellation would consist of 840 satellites plus four spares at 700 km altitudes in each of 21 orbital planes in near-polar inclinations. By using a 98.2 degree, sun-synchronous inclination, the satellites' solar cells remain aligned with the sun. The number of satellites is large enough to reduce occurrences of possible signal attenuation due to atmospheric moisture. The satellites will operate in the 20 GHz/30 GHz bands which are susceptible to this phenomena." The antenna's footprint will be limited to a radius of about 700 km, requiring a large number of satellites to provide seamless global coverage. The higher frequencies allow 20,000 simultaneous digital connections, each carrying 1.54 million bits per second of video or data.

Teledesic expects to be able to build all of the satellites by the year 2000 and have them all in place by the end of 2001. Profits the first year are projected to be more than a half-billion dollars, and at the end of the fifth year, profits should reach $6.5 billion. In the *New York Times*, Russell Daggatt, president of Teledesic, based in Kirkland, Washington, said, "The McCaw-Gates system would complement rather than rival cellular telephone networks. Unlike the Iridium or Loral systems, it will go beyond just telephone service and be able to relay digital medical images, bulky computer data files, and two-way video conferences. These types of signals are packed with so much electronic information that they require the high-capacity, or 'broad-band' communication pathways Teledesic plans to provide." Normally, wireline communications are needed to send that type of data. The system will also be capable of handling 100,000 simultaneous conversations.

Analysts have mixed reactions to such an ambitious proposal:

An article in Business Week states, "But each one [satellite] can only provide service to a circle 450 miles in radius. That means many satellites are needed to blanket the globe—and without global coverage, the system probably will never generate the traffic to be economical. Says former FCC lawyer Richard M. Firestone: 'If the costs are too high, they won't get the volume they need, but they can't get the costs down unless they get the volume. It's a vicious cycle.'"

"I make the analogy to the computer industry that evolved from big mainframes," Daggett says. "If you were thinking of sending up 840 of those powerful mainframes, people would think you were crazy. But if you were thinking of sending up 840 notebook computers, it is a different matter altogether."

12.4.2 Iridium

According to *Financial World*, Motorola's motives for entering the Low Earth Orbit satellite race with Iridium are a result of the limitations of the conventional cellular systems: too little capacity and too much interference. Cellular phones account for 30 percent of Motorola's sales and 35 percent of their earnings. In addition, competition is causing the price of phones to drop. Rather than wait for improvements in land-based cellular systems, they are making the jump to satellite communications with Iridium.

Some of their partners include Sprint, BDE Canada, Italy's telephone holding company, Stet and, in Japan, Daini Dended, Mitsubishi, and Kyocera. The satellites will be built by Lockheed with subcontractors Raytheon, Scientific-Atlanta, Martin Marietta, and Siemens A.G. Sprint and BCE Canada will build and control the North American gateway. The system will consist of 66 satellites that will cost $13 million each. As of April 1994, they have $1 billion toward the estimated total cost of $3.37 billion. The satellites will be in 420 nautical mile orbits, radiating 48 spot beams, with 11 satellites in each of six orbital planes. They plan to use TDMA technology and plan to be operational by 1998.

As stated in *Financial World*, Motorola plans to give a bit of the revenue from each phone call to every national telephone system it bypasses. According to *Aviation Week and Space Technology*, they plan to use the satellites as switching stations and connect to the nearest earth station of the party being called, therefore "bypassing en route public telephone service providers except when actually connecting with the number being called." The estimated cost of a phone call will be $2 to $3 per minute. Analysts believe their target market will be "high-end users such as executives." A company spokesperson, John Windolph, claims that their system will have a higher signal strength than Globalstar's, thus making it more appealing.

12.4.3 Globalstar

An article in *Financial World* [18] explains how Globalstar is expected to work: "For example, a call traveling from a desktop office telephone in New York City to a Globalstar customer wandering in the Sahara would travel through traditional means until the signal reached a gateway, perhaps in Cairo. That's when it would be bounced off the satellite down to the customer's handset."

This system is primarily a joint venture between Loral and Qualcomm. As of April 1994, they had a commitment of $275 million of the estimated $1.8 billion cost from their other partners, which will be enough for the first phase of the system. The partners, who are primarily telecommunications companies, are expected to have roles as providers of Globalstar's mobile phone service. Loral chairman, Bernard Schwartz, said, "nearly all of the Globalstar investors provide communications services themselves and will help in delivery of Globalstar service in their areas. That will speed local approvals and hold down costs." So far, ten of the partners plan to offer service in 33 countries: 14 in Europe, 8 in Asia, 5 in Africa, and 6 in North/South America. Some of the partners include: France's Alcatel-Alsthom, Korea's DaCom and Hyundai Electronics, Deutsche Aerospace, Britain's Vodafone, San Francisco's AirTouch Communications, Qualcomm, Space Systems/Loral, and Italy's Alenia Spazio.

The first 24 of 48 satellites are expected to be in place by 1998, with the remainder in place within the following year. Because of its partnership with Qualcomm, Globalstar will use CDMA technology for its transmissions. According to the *Wall Street Journal*, [28] "By 2002, they anticipate having 2.7 million subscribers and annual revenue of $1.6 billion, and by 2012, as many as 16 million users worldwide are projected."

Analysts say that the service will appeal more to "far-flung governments and communities seeking lower-cost service." The primary market will be to "wholesale mobile communications services to distributors such as existing telecommunications companies and government ministries. Direct satellite service would be offered where ground-based telecommunications systems are lacking, such as in remote and underdeveloped countries." The cost of the service is projected at $0.65 a minute. The service will provide more revenue for foreign governments than Iridium because of how the signal will travel to the user. In addition, a *New York Times* article states, "By using existing local ground networks to beam calls to an international satellite network, the Globalstar system will be able to use fewer and less complex satellites than other mobile communications systems."

12.4.4 Inmarsat

Inmarsat, the International Maritime Satellite Organization, based in London, with members from 70 nations, has the benefit of being the only organization offering mobile phone service on the market today. According to *Aviation Week and Space Technology*, they currently have more than 31,000 terminals in use. They offer three types of mobile phone service. The oldest system, an analog FM Inmarsat-A phone, is being replaced by Inmarsat-B, a digital system. The newest system, Inmarsat-M, sacrifices a bit of voice quality/data rate for a reduction in the weight, size and cost of the equipment. The cost ranges from $15,000 to $40,000. Their fourth service, Inmarsat-C, which is a data-only message service, has equipment prices of about $4,500. "Customers may choose not only among equipment manufacturers but also among service providers."

Again, according to an *Aviation Week and Space Technology* article, Inmarsat has dismissed using LEOs and is considering an intermediate circular orbit or geosynchronous satellite for their next mobile phone venture. The satellites that they have up now feature "the first commercial application of spot beams in the L-Band and are switchable between spot and global in any proportion of total power. Inmarsat hopes the spot beams and a very agile frequency reallocation scheme will add up to the exceptional operating flexibility through 2008, the satellites' design lifetime."

Currently, in the airline industry, three consortia provide passengers with data and voice services based on the Inmarsat system. As of September 1993, aeronautical voice and low-gain data systems had been commissioned on over 300 aircraft worldwide, including 22 airlines, with many more placing orders.

An advantage the Inmarsat system may have over the competition will be its ability to obtain financial support and international approval more easily due to the nature of its organization. In some ways, the organization of the Inmarsat charter could also prove to be a disadvantage. Each member's contribution is based on the previous year's use of the currently available assets; for example, Comsat Corp has a 22 percent share toward contributions whereas British Telecom has 11 percent. Comsat may be reluctant to contribute so much money because of other competition present in the U.S., but other nations that will benefit more from such a system may wish to contribute more than their required share. To allow for that situation, the Inmarsat charter would have to be changed. The previous change of such magnitude took five years. If this were to happen before work could proceed on their new system, they may enter the race too late to compete.

12.4.5 Ellipsat

Ellipsat, which is a subsidiary of Mobile Communications Holdings, Inc., has a unique design for its constellation of satellites. It plans to provide peak system capacity during daylight hours to the areas with large populations. According to an article in *Aviation Week and Space Technology*, the satellite design will be done by Fairchild Space and Defense Corporation. Harris Corporation will handle the communications payload, with Israel Aircraft Industries as a subcontractor. Westinghouse, also an investor, will be handling the earth terminals/system integration. Their southern hemisphere partner is Carincross Holdings Pty. Ltd., of Sydney, Australia.

To implement Ellipsat, they will use two different types of orbits. The first orbit, Borealis, will be an inclined elliptical orbit with an apogee of 7800 km, a perigee of 520 km and an orbital period of 3 hours. The apogee will be located in the northern hemisphere to provide a relatively long-duration visibility north of the equator. The second orbit, Concordia, will have an apogee of 7800 km and a perigee of 4000 km, which will provide an equatorial elliptical orbit. The orbital period will be 3 to 4 hours, with the main service in regions south of the equator with latitudes of approximately 40 degrees and some service to mid-latitudes north of the equator.

The assumption in this system is that peak demand is during daylight hours. The satellites in sun-synchronous inclinations of 116 degrees will enable the peak capacity to follow the sun and therefore the demand. Both use the same satellites; "with only four satellites deployed in the inclined Borealis orbits, we can obtain daylight service in the northern hemisphere," according to David Castiel (president of MCHI). "With the addition of four more satellites, 24-hour coverage is provided. And with four additional satellites in equatorial orbit, we can provide daylight service in the mid-latitudes both south and north of the equator. With another four satellites in equatorial orbit—a total of 16—we extend mid-latitude coverage south of the equator to 24 hours." They already have FCC approval to launch six experimental satellites.

12.4.6 Orbcomm

Orbcomm is the Orbital Sciences Corporation subsidiary that was "formed to bring the world's first space-based mobile data communications system to market," according to an *Aviation Week and Space Technology* article. Analysts believe that because it is the farthest along, targeting a niche market, and has a low cost, it may have the best chance of succeeding. In addition, its main international partner is Canadian Teleglobe of Montreal, Canada. According to the *Washington Times*, [17] "Teleglobe is the Canadian signatory to the Intelsat and Inmarsat treaties, making it the exclusive provider of satellite services to make international calls to and from Canada, except those to and from the United States."

The company will launch two initial satellites plus 24 more launched in three groups of eight. The system will initially have four Orbcomm gateway earth stations in Arcade, NY, Arizona, Washington, and Georgia, with the control center in Dulles, VA. Kyushu Matsushita Electric has designed two types of operational subscriber terminals. Vito/Net, a full-feature unit, is equipped with a seven-line screen and alphanumeric keypad to send and receive messages similar to electronic mail. The other unit is equipped with a computer interface port, and is designed to provide an emergency service called SecureNet and data acquisition and monitoring services offered by DataNet." The communicators may sell for as little as $50.

Orbcomm U.S. will be the primary marketer of the communication services in the U.S., and Orbcomm International will establish and manage licensee-operated Orbcomm networks elsewhere. As described in the *Wall Street Journal*, IDB Mobile Communications will market the Orbcomm services to marine industries including shipping, fishing fleets, oil rigs and tankers.

12.4.7 Odyssey

Odyssey is being proposed by TRW and is unique in that it is being designed with medium altitude orbits. The system will comprise 12 satellites in 6,400-mile medium earth orbits with 55 degrees inclination. Each satellite will have a six-hour period in three orbital planes. According to *Aviation Week and Space Technology*, [32] "Each satellite would remain in view of a mobile user for about two hours, minimizing the need to transfer to another satellite. The higher orbital altitude also would provide high latitude coverage and enable two gateway terminals to serve an area the size of the continental U.S." The satellites will have the ability to reorient their multibeam antennas toward areas of high population concentrations during peak hours. It is estimated that each handset will need to radiate only 0.5 watt average power, and the system will be able to handle three to six million subscribers.

12.4.8 Aries/Constellation

Constellation Communications, Inc. is developing the Constellation satellite system, which was formerly named Aries. They plan to place 48 satellites in inclined orbits at 2000 km altitude. This altitude, which is twice what Globalstar is proposing, "is to assure that a handheld telephone user need not communicate with a satellite at a viewing angle of less than 15 degrees," according to Bruce Kraselski, CCI senior vice president. The system will be able to handle more than 1000 voice channels and is expected to cost around $1.2 billion.

12.5 CONCLUSION

Challenges exist to convince the U.S. government and foreign governments to sanction LEO efforts to attract and convince financiers that returns outweigh risks in this unproved market. Several systems are pursuing similar markets and similar sources of funding; only time will tell who the winner is. The potential market size will probably be greater than most estimates. As stated in the *New York Times*, "While supply often creates its own demand, it is impossible to predict what people will do if a cornucopia of new services becomes available." When automobiles first came on the market, they were thought of as an expensive toy. Now, a modern-day society cannot function without them, and most people spend a great deal of time in their cars. Similarly, the cellular phone market continues to grow, and what was once thought of as an expensive toy—the cellular phone—is gradually becoming a necessity for drivers. The growth of international business will provide one market for the satellite-based communication systems; as the world becomes smaller, more people will look upon these wireless systems as a necessity.

The challenges facing the systems are both nontechnical and technical in nature. The technical ones do not seem insurmountable and government intervention will probably cause more headaches for the corporations involved. The nontechnical ones, which include licensing time, and deal mainly with deploying and managing systems that are miles overhead.

Of the systems outlined, Inmarsat is the only one with an operational satellite-based phone service. Its market base is small because the cost of the units is so high. To keep up with the competition, Inmarsat will need to implement a more cost-effective system. Orbcomm will probably be the next system to come on line, according to industry analysts. Their partner, Canadian Teleglove, is a member of Inmarsat and will probably not want to invest in any new Inmarsat system, thus hurting its chances. Teledisc's system has the largest number of hurdles to overcome because of its enormity. The manufacturing challenges alone will take years to work out. Management of a system with 840 satellites will force them to develop innovative solutions, and they may not meet the prediction of being operational by 2001, which is less than 5 years away. Iridium and Globalstar seem to be the two closest competitors by planning to be operational by 1998. Their biggest challenge will probably be in obtaining licenses from foreign governments. Globalstar should have the advantage by not bypassing local phone services, by which foreign governments usually obtain revenue. Ellipsat may be hampered by the fact that they make the assumption that peak user hours are during daylight only. Many cultures do not "roll up the sidewalks and go to bed" when the sun goes down. Often, many phone calls and socializing are done during the late evening. In addition, when it is night in Japan, for example, business is just starting in the U.S. If business conversations must take place, odds are it will be night in one of the two locations. Odyssey and Aries are keeping very quiet about their systems, and only time will tell whether they succeed or fail.

Time will also tell how successful and profitable these systems become. Maybe by the 21st century all of the world's citizens will have their own personal phone numbers that will travel with them wherever they go.

REFERENCES

[1]   Ha, Tri T. *Digital Satellite Communications*. New York: Macmillan, 1986.

[2]   Pratt, Timothy, and Charles W. Bostian. *Satellite Communications Systems*. (New York: John Wiley and Sons, 1986.)

[3]   Maral, G., and M. Bousquet. *Satellite Communications Systems*. (Chichester, England: John Wiley, 1984.)

[4] Couch, Leon W. *Digital and Analog Communications Systems.* (New York: Macmillan, 1990.)

[5] Pritchard, Wilbur L., and Joseph A. Sciulli. *Satellite Communications Systems Engineering.* (Englewood Cliffs, New Jersey: Prentice-Hall, 1986.)

[6] Ziemer, Rodger E., and Peterson, Roger L. *Digital Communications and Spread Spectrum Systems.* (New York: Macmillan, 1985).

[7] Lin, S. and Costello, D.J. Jr. *Digital and Analog Communication Systems. (Englewood Cliffs, New Jersey: Prentice-Hall, 1983).*

[8] Van Tress, Harry L., Ed. *Satellite Communications Systems.* (New York: IEEE Press, 1979).

[9] Forney, G. David Jr. "The Viterbi Algorithm" *Proceedings of the IEEE*, 61: 268-278, March 1973.

[10] Heller, Jerrold A. and Irwin Mark Jacobs. "Viterbi Decoding for Satellite and Space Communications." *IEEE Transactions on Communications* COM-19: 835-848, October 1971.

[11] Andrews, Edmund L. "The New Space Race in Satellite Communications." *The New York Times*, March 27, 1994: 3, 9:1.

[12] Asker, James R. "Inmarsat Braces for Multiple Rivals." *Aviation Week and Space Technology*, October 11, 1993: 139 (15):44-45.

[13] Cole, Jeff. "Loral Funds Start of Global Phone System." *The Wall Street Journal*, March 24, 1994: A, 4:1.

[14] "Phones into Orbit." *The Economist*, March 28, 1992, 332 (7752) 14–18.

[15] "Financing for Global Phone System Is Set." *The New York Times*, March 25, 1994: D:33.

[16] Fox, Barry. "Spymasters Fear Bug-proof Cellphones." *New Scientist*, January 30, 1993: 137 (1858): 19.

[17] Gibbons, Kent. "Oribital Signs Canadian Partner for Satellite Mobile Message Effort." *The Washington Times*, April 21, 1993: C, 3:2.

[18] Hass, Nancy. "Preemptive Strike." *Financial World*, September 14, 1993: 162 (18): 36-39.

[19] "In Brief," *Broadcasting*, August 10, 1992: 122 (33): 65.

[20] Jaffe, Thomas. "Hillary Clinton, Smart Investor," *Forbes*, May 10, 1993: 151 (10):18.

[21] Klass, Philip J., "New Investors, Contenders Vie for Low-Earth Orbit Service," *Aviation Week and Space Technology*, April 4, 1994: 140: 57-58.

[22] Klass, Philip J. "WARC-92 Approves Satellites for Small Cellular Telephones," *Aviation Week and Space Technology*, March 9, 1992: 136 (10):31.

[23] Klass, Philip J. "Low-Earth Orbit Communications Satellites Compete for Investors and U.S. Approval," *Aviation Week and Space Technology,* May 18, 1992: 136 (20):60-61.

[24] Lavitt, Michael O. "Industry Outlook," *Aviation Week and Space Technology*, December 14/21, 1992: 137 (24):13.

[25] Lewyn, Mark. "He's No Mere Satellite-Gazer," *Business Week*, April 4, 1994: (B365): 39.

[26] McKenna, James T. "LEO Endeavors Face Finance Hurdles, Red Tape," *Aviation Week and Space Technology*, June 7, 1993: 136 (20): 112-113.

[27] Nordwall, Bruce D. "Filter Center," *Aviation Week and Space Technology*, September 21, 1992: 137 (12): 55.

[28] Orbital Sciences Marketing Pact," *The Wall Street Journal*, February 10, 1994: A3:3.

[29] Stix, Gary. "Cyberspace Cadets," *Scientific American*, June 1994: 270 (6): 98-101.

[30] Sugawara, Sandra. "Satellite Network Seeks to Link Remote Areas," *The Washington Post*, March 21, 1994: A7:1.

[31] "Teledesic Pushes $9-Billion, 900-Satellite System," *Aviation Week and Space Technology*, March 28, 1994: 140 (13): 26.

[32] Velocci Jr., Anthony L. "Full Funding Allows OSC to Retain Orbcomm Unit," *Aviation Week and Space Technology*, August 16, 1993: 139 (7): 74-75.

Chapter 13

Leo Signal Processing Design for Telestar I

The Telestar I Satellite System was conceived by a team of Virginia Tech satellite engineers working under Dr. Fred Ricci, as a complement to the Orbcomm System. Telestar provides mobile satellite data communications for a part of the original Orbcomm market should funding, technology or luck work against Orbcomm development.[1] Telestar will include the latest digital communications and networking technology to substantially improve the Orbcomm proposed services for electronic mail (e-mail) and emergency aid requests. When Telestar's e-mail and emergency aid services produce positive results (and profits), further development of additional services will be undertaken. As a business decision, further development of the Telestar system through expanding to more services and customers will be a logical and rational risk. E-mail and emergency messages are only a part of the original Orbcomm system that proposes to support position tracking of truck-trailer shipping, data monitoring and control of power production networks, personal paging, and marine message and tracking services as well. Telestar incorporates "lessons learned" from scrutiny of the original Orbcomm FCC filing and includes the latest satel-

---

[1]Launched on April 12, 1995, the first two Orbcomm satellites developed immediate problems with GPS position communications and satellite command and control, proving that a fallback design such as Telestar I is an alternative.

lite technology and sophisticated design to optimize performance, maximize throughput and minimize cost and risk.

13.0 THE TELESTAR SIGNAL PROCESSING FUNCTIONS

The Telestar project consisted of teams of engineers working on the following specific systems: license and finance, launching, satellite payload, earth terminals, satellite control, and system architecture. The satellite payload system is further subdivided into antennas, signal processing, computers and telemetry, tracking and command. This section defines the signal processing functions of the Telestar system, herein referred to as "the system." The signal processing functions are critical to the success of the project, since they implement the actual transmitted waveform that will carry data back and forth to the Telestar satellites. The signal processing functions also implement the multiple access technique that allows the many Telestar users to have equal access to satellite services. Finally, the signal processing functions incorporate the error checking and correction that will ensure clean, consistently accurate transmission and reception, a key selling point and crucial to the performance of the Telestar system.

13.1 SIGNAL PROCESSING SUBSYSTEM REQUIREMENTS

The Signal Processing functions of the Telestar Satellite System must optimize system performance for the minimum cost and complexity. Evaluation of the system requirements will use the design methodology shown in the appendix to this chapter. The subsystem requirements for Telestar signal processing are listed as follows:

Radio Regulations and Spectrum

1. Utilize the Federal Communications Commission (FCC) allocations of bandwidth in accordance with the World Administrative Radio Conference (WARC) approved mobile satellite service spectrum. Telestar will use the following VHF and UHF frequency bands for the listed uses.[2]
   - 137.00–138.00 MHz satellite downlink to earth station subscriber communicators (SCs) and Gateway Earth Stations (GESs)
   - 148.00–150.05 MHz satellite receive uplinks from SCs and GESs
   - 400.05–400.15 MHz satellite GPS-derived Standard Frequency and Time signal. (downlink)

Service Requirements

2. Support low-cost two-way mobile and portable e-mail messaging and emergency aid messaging for the primary Telestar coverage areas. These coverage areas include loca-

---

[2]The frequency bands listed are the maximum allocated and approved frequency bands.

tions throughout the continental U.S. (CONUS) and central and western Europe. Other areas of the earth are partially covered for fractions of each day as a result of the constellation design. The system architecture and space launch documents describe this part of the system. E-mail messages are transmitted to these same locations via the Telestar system, or other locations via Internet, Public-Switched Telephone Network (PSTN), and other ground-based systems through GESs.[3]

Terminal Design: Terrestrial and Satellite Links

3. Minimize the effective isotropic radiated power (EIRP) and digital signal processing requirements of the SCs through satellite design techniques incorporating:

- Modulation mode selection for the minimum ratio of energy per bit to noise power spectral density (E_b/N_0)
- Multiple-access control for contention resolution and maximum throughput
- Signal encoding and error correction for minimum undetected errors even in fading and weather-affected propagation

Traffic Assumptions: Terminal Activity

4. Support automated emergency aid request messages in the Telestar primary coverage areas. Automated position location is provided and a "return acknowledge" path is also to be available to confirm receipt of emergency messages.

5. Place priority on emergency aid messages over e-mail messages, to ensure that life-threatening situations are handled before routine message traffic.

Payload Dimensions: On-board Processing

6. Minimize satellite transmission delays due to payload processing. Satellite processing delays will slow the system and decrease overall capacity.

7. Minimize payload complexity in order to minimize the weight of signal processing equipment. Complexities that add equipment weight have an impact on launch costs and system costs. In addition, a less complex payload design is likely to enhance problem resolution in the early operational testing phase of the Telestar program. (Note that we will address the complexity issue without distinctly analyzing the exact weight of the equipment considered). The Telestar space launch team will produce a true satellite weight budget prior to a full-scale production decision.

Networking

8. Maximize the number of users on the Telestar system by an intelligent choice of a multiple-access scheme. Related to this requirement is the need to maximize both the numbers of simultaneous users on a single satellite and the total users of the system over time.

---

[3]Traffic load assumptions are part of the detailed system architecture. This analysis uses 350,000 subscribers sending two new message requests per day.

Link Quality

9. Minimize the number of errors occurring per transmission and increase the accuracy of received transmission by an optimum choice of an error detection and correction method for communications.

13.2 DEFINING THE FUNCTIONS OF TELESTAR SATELLITE SIGNAL PROCESSING

The functions to be included in the signal processing subsystem are defined as the digital signal processing functions that produce baseband information in the bandpass channel at the payload. [1] The Telestar satellites will be involved in many different functions to support communications including telemetry, tracking and command, attitude and orbit control, power distribution and antenna control. The major functions of signal processing include all the digital satellite functions occurring on the satellite between the uplink receive intermediate frequency (IF) amplifier and the downlink transmit IF amplifier, as shown in Figure 13.1

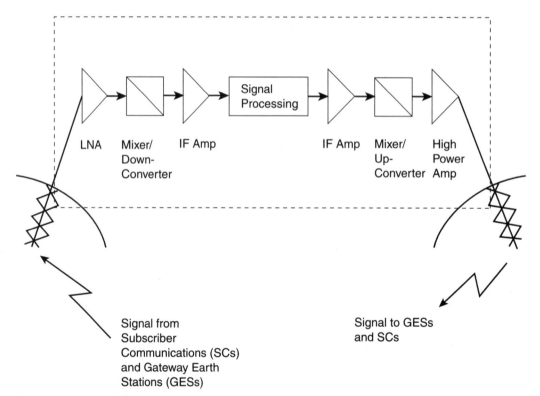

Figure 13.1 Payload Block Diagram.

The choices regarding the specific functions of Telestar signal processing satisfy the major requirements of minimizing processing delays and complexity and maximizing user capacity and message accuracy. Specifically, the following signal processing functions are defined:

- Modulation Mode
- Multiple Access Scheme
- Message Format
- Error Detection and Correction

These signal processing functions appear on the payload as shown in Figure 13.2.

The signal processing modulation functions are implemented in the demodulator and modulator functions shown in Figure 13.2. The multiple-access scheme is implemented by circuitry in the payload that decodes the bit-stream from the demodulator and determines whether accurate data was received. Accurate data is then stripped of its error-correction and header data, corrected if possible, and routed to a correct downlink channel.

Payload Dimensioning: Payload dimensioning includes the number of satellite channels, total RF power, satellite antenna sizes, and the satellite amplifier power. The choice of RF modulation scheme is in some ways limited by the original scope of the Orbcomm project, in that the modulation modes, frequency plan and channelization have all been defined. Table 13.1 lists these parameters as understood for the Orbcomm satellite system.

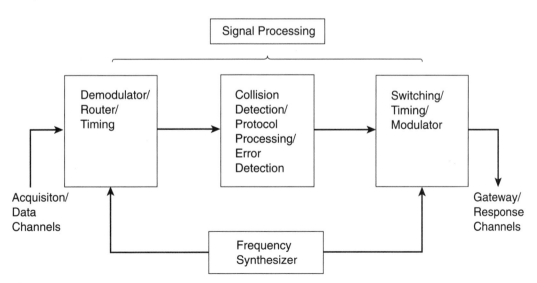

Figure 13.2 Signal Processing Functions.

Table 13.1 Orbcomm FCC Filing Signal Processing Parameters

	Orbcomm Signal Processing Parameters[1]				
	SC to Satellite (Uplink)	Satellite to SC (Downlink)	GES to Satellite (uplink)	Satellite to GES (Downlink)	Standard Time and Frequency
Frequency Band (MHz)	148.00- 148.378	137.00- 137.270	148.700- 148.800	137.300- 137.400	400.075- 400.125
Data Rates (kbps)	1.200	4.800	57.6	57.6	4.800
Modulation Mode	FSK	PSK	PSK	PSK	NIST[2]
Number of Channels	20	8	1	1	1

[1]Does not include the GPS frequency plan

[2]The use of this channel is to be coordinated with the National Institute for Standards and Technology (NIST) to provide unrestricted public use of time signals accurate to 1 μs.

The SC uplink modulation mode and data rate used in the current Orbcomm system differs from that shown in Table 13.1. A similar change for the Telestar system is discussed below.

13.3 MODIFICATIONS TO THE FCC FILING

Modulation Modes: The Telestar system will implement all satellite links by using phase shift key (PSK) modulation. The Telestar SC uplink will be a PSK link at 2400 bps, versus the Orbcomm FCC filing data specifying a frequency shift key (FSK) modulated link at 1200 bps. The improvement in data rate to 2400 bps is possible from the spectral efficiency gains that PSK modulation contributes over FSK modulation. [2] Increasing the system spectral efficiency (that is, capacity in bits-per-second-per-Hertz) allows a higher data rate and throughput and predictably improves the system performance. PSK modulation also has the benefit of a constant amplitude envelope, providing a fixed-carrier power signal over time and making it particularly suited to satellite communications.

PSK Constant Amplitude Envelope: Most satellite systems use phase modulation for digital transmission because of the constant amplitude characteristics. With PSK modulation, only the signal phase changes from data symbol to symbol. Equation 13.1 shows a PSK signal, s(t) in terms of a sinusoidal function with angular frequency $2\Pi f_c$ and phase $\theta_c = ui(^{\Pi}/2)$.

$$s(t) = V\cos(2\,\Pi f_c t - u_i(^{\Pi}/2)) \tag{13.1}$$

$$u_i = \pm1, +1 = \text{logical } 0, -1 = \text{logical } 1$$

$$s(t) = Vu_i\sin(2\Pi_c t) \tag{13.2}$$

BPSK modulation sets the carrier phase θ_c equal to $-\Pi/2$ (-90°) to indicate a binary 0 and $+\Pi/2$ (+90°) to indicate a binary 1. The resulting carrier amplitude envelope does not fluctuate. Equation 13.2 is another form of equation 13.1, derived using trigonometric identities. BPSK modulation thus resembles amplitude modulation, where the carrier power is stable with amplitude ± V. This stable amplitude allows a satellite receiver to decode a fixed carrier-to-noise (C/N) ratio under additive white-Gaussian noise conditions. Stable carrier amplitude also simplifies filtering, providing a clean intermediate frequency (IF) signal for the satellite receiver's bandwidth noise filters.

BER Improvement: The C/N at the satellite receiver demodulator directly affects the signal bit-error-rate (BER). The Telestar system performance value for BER is 10^{-6}. Given a fixed BER, PSK modulation improves link quality over FSK modulation. Figure 13.3 shows a 3dB advantage of using PSK over FSK at 10^{-6}. FSK requires 3 dB more

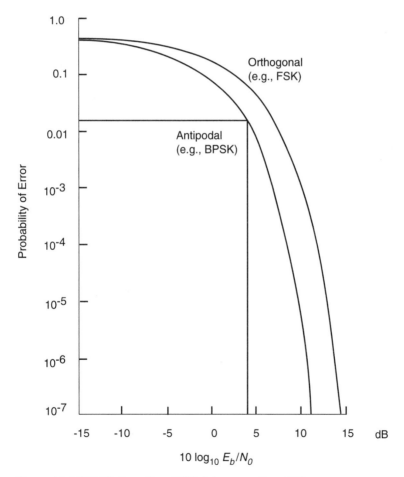

Figure 13.3 PSK Bit-Error Rate (BER) Advantage Over FSK.

power than BPSK for the same BER. PSK has an additional advantage over FSK for the satellite transmitters, since the transmitter can operate at saturation for maximum power efficiency.

Power Spectral Density: PSK encoding (modulation) also improves the system spectral occupancy over FSK as measured by spectral efficiency. Spectral efficiency is defined as the ratio of the capacity R_c (bps) to the bandwidth B (Hz) of a single modulated carrier. The maximum theoretical spectral efficiency of BPSK modulation is 1 bps/Hz, since each data bit represents a signal transition. The actual spectral efficiency degrades due to nonoptimal filtering and nonlinearities to a range between 0.7 and 0.8 bps/Hz. The theoretical and actual spectral efficiency for FSK will be less than these PSK values. [3]

Partially Coherent Differential PSK: To satisfy the Telestar signal processing requirements for minimal payload complexity, Telestar will have to use partially coherent differential PSK (DPSK). A *coherent* modulated signal requires two inputs, one for a reference signal like a synchronized oscillator and one for the modulated signal. [4] Since PSK requires a reference that is phase coherent with the transmitter, PSK modulation must be coherent. Coherent detection, however, would force the Telestar signal processing to include carrier recovery circuitry, such as phase lock loop processing. As an alternative to this additional processing, PSK modulation will minimize the payload processing while gaining spectrum and BER advantages.

PSK modulation is bandpass modulation, requiring a mixer, an IF amplifier and a detector. The product detector in PSK modulation acts like a phase detector whose output is directly proportional to the digital input of the baseband signal. BPSK modulation by definition must be coherent, since there must be a reference voltage that has the same phase as the original unmodulated carrier. However, DPSK provides a partially coherent technique that uses a delayed version of the transmitted signal as a reference. With partially coherent DPSK modulation, no additional reference is required. Equations 13.1 and 13.2 are modified for DPSK, so that a logical 1 would result from no phase changes between PSK symbols, and a logical 0 would signify a phase reversal. DPSK in practice might require, at most, 1 dB more in E_b/N_0, but it more than makes up for this in payload simplicity. DPSK demodulation is shown in Figure 13.4.

BPSK Versus Quadrature PSK: Telestar will use BPSK modulation over quadrature PSK (QPSK) modulation because BPSK increases the simplicity of the payload design while increasing the accuracy of the Telestar data links. QPSK takes two bits at a time to define one of four symbols, corresponding to one of four carrier phases. The QPSK symbol rate is one half the BPSK rate and has the accompanying advantage of half the spectrum for the same user information. This spectrum advantage comes at the cost of higher BER and twice as much signal processing on the payload. This is because QPSK is equivalent to *two* BPSK signals, one using a cosine carrier and the other using a sine carrier. A four-state phase modulator is shown in Figure 13.5.

BPSK, shown in Figure 13.6, is more accurate than QPSK, given the same data rate Rc, bandwidth B, and C/N ratio [2]. In addition, QPSK is subject to greater phase unbalance degradation than BPSK, since there are twice as many phase states and phase transitions. When large phase transitions are felt at the satellite receiver filter components,

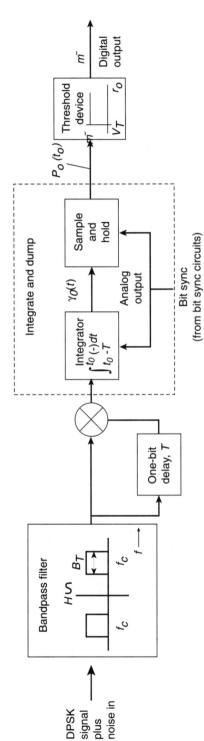

(a) A Suboptimum Demodulator Using an Integrate and Dump Filter

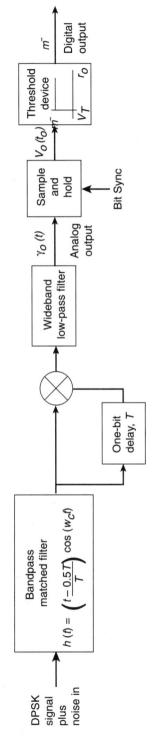

(b) A Optimum Demodulator Using a Bandpass Matched Filter

Figure 13.4 DPSK Demodulation.

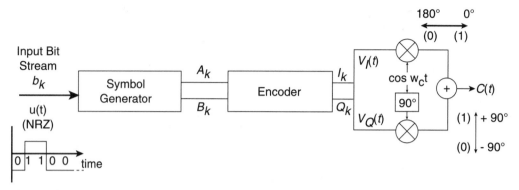

Figure 13.5 Four-State QPSK Phase Modulator

amplitude modulation of the carrier may result [3]. The nonlinearity of the filtering transforms these variations into phase changes that degrade the receiver performance. Figure 13.7 shows the degradation due to phase unbalances in a QPSK modulator at 10^{-6} BER.

A BPSK modulation scheme for Telestar ensures that the system utilizes the available bandwidth wiht the minimum payload complexity. With a higher-power spectral density than FSK modulation, BPSK still allows Telestar to implement a medium-access scheme for many users using a band width that is significantly smaller than similar systems.

13.4 MULTIPLE-ACCESS SCHEME AND MESSAGE STRUCTURE

In order to determine an appropriate multiple-access scheme, a complete understanding of the multiple access requirements is imperative. A comparison of multiple-access techniques will determine which is best for Telestar given its unique requirements.

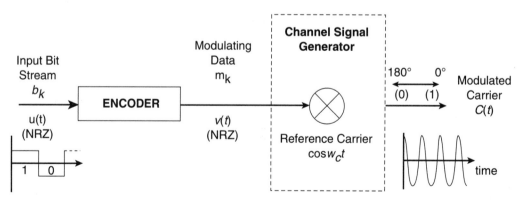

Figure 13.6 BPSK Phase Modulator

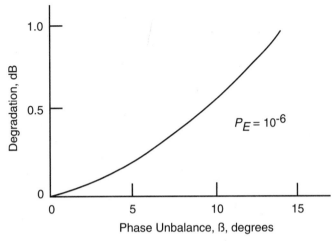

Figure 13.7 Degradation Due to Phase Unbalance in QPSK Modulator

Requirements: Telestar must be able to accommodate many users with short, infrequent, random transmissions. In addition, users can send different types of messages—emergency or electronic mail (e-mail). Emergency messages are the equivalent of a call for help and should contain the minimum amount of information necessary to enable rescuers to find the distressed subscriber. E-mail messages will be much longer, up to 240 characters long. The FCC proposal defines some requirements, such as the number and frequency of subscriber uplink and downlink channels. In summary, the multiple access requirements for Telestar are:

1. Accommodate variable message types.
   - Emergency message (EM)—infrequent, random, short (few characters)
   - E-mail—infrequent, random, variable length (0–240 characters)
2. Provide access to several hundred thousand users.
3. Subscriber uplink bandwidth is 378 kHz.
   - Twenty 15-kHz subscriber uplink channels (FDMA)
4. Subscriber downlink bandwidth is 270 kHz.
   - Eight 27-kHz subscriber downlink channels (FDMA)

Pure Frequency Division Multiple Access (FDMA) would provide each subscriber a single uplink and downlink channel. Given the frequency plan defined in the FCC proposal of 20 subscriber uplink channels and 8 subscriber downlink channels, pure FDMA is clearly insufficient to support several hundred thousand subscribers. Therefore, in order to fulfill the requirements of providing access to several hundred thousand users, Telestar must implement another multiple-access scheme in addition to the frequency channels provided.

13.4.1 Comparison of Multiple-Access Techniques

The multiple-access techniques available to combine with Telestar's FDMA scheme are Time Division Multiple Access (TDMA), Code Division Multiple Access (CDMA), Random Access, and Demand Assigned Multiple Access (DAMA).

TDMA: In TDMA, each subscriber would have an assigned time slot during which to transmit messages. In other words, subscribers would take turns transmitting messages. With hundreds of thousands of subscribers, the resulting time between each subscriber's time slot would be very long, on the order of minutes to hours, depending on the number and duration of time slots. Such a system would be impractical to users, especially those who need to send an emergency message. Also, because the frequency of transmissions over time will be very low, TDMA would be extremely inefficient with most time slots remaining unused.

CDMA: In CDMA, each subscriber terminal spreads its transmissions over the entire carrier bandwidth through the use of a binary sequence, or code. CDMA allows multiple users to transmit continuously and simultaneously on the same frequency channel. [3] At first glance, CDMA might be an appropriate choice for Telestar; however, the trade-off of continuous, simultaneous transmissions is a low throughput of around 10 percent as shown in Figure 13.8. [3] Telestar terminals do not require the capability to transmit continuously, thus allowing the system to implement a multiple-access scheme with a greater throughput.

Random Access: Random access is very effective in systems similar to Telestar that experience short, infrequent, random transmissions. A common random access technique is Aloha random access, which allows users to transmit messages at any time. After a user transmits a message using the Aloha protocol, the satellite sends an acknowledgment (ACK) back to the user, indicating a successful transmission. If two users attempt to transmit and their messages overlap in time, the satellite cannot process either message and therefore cannot send an ACK to the terminal. These overlapping messages are called collisions. If terminals send messages that collide, they wait for different amounts of time before attempting to transmit again. A terminal will continue to transmit its message until it receives an ACK from the satellite. Telestar's requirement to accommodate messages with varying lengths makes Aloha inappropriate because transmission of the longer e-mail messages would increase the probability of message overlap and result in too many collisions.

DAMA: In demand assigned multiple access, users request frequency channels and/or time slots to transmit messages. The amount of resources user terminals request depends on the length of the messages they must transmit. Because DAMA accommodates messages with varying lengths, further analysis shows it is most appropriate for Telestar.

13.4.2 Design of DAMA for Telestar

DAMA requires a means for terminals to request resources from satellites and satellites to assign resources to terminals. Of the 20 uplink frequency channels, some will act as request channels and others as assignment channels. Terminals send a request for resources to the

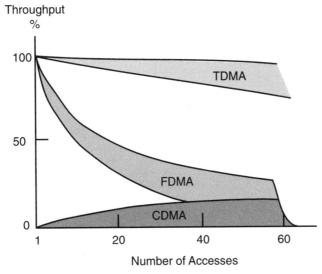

Figure 13.8 Throughput Comparison
Source: Reprinted from [3].

satellite on one of the request channels. The satellite responds by assigning to the terminal a frequency channel and time slots in which the terminal will transmit its message. Terminals will transmit emergency messages on a request channel. A flow diagram is presented followed by a complete explanation of the DAMA protocol for Telestar in Figure 13.9.

If a terminal is transmitting an emergency or e-mail message, it transmits a request on one of the request channels. If the terminal does not receive an acknowledgment within eight time slots after it transmits the request, it assumes a collision occurred and transmits again. To ensure that colliding messages do not continue to collide indefinitely, each terminal after detecting a collision will wait a random number of time slots and randomly retransmit on a different request channel. To give emergency messages a higher priority than e-mail requests and to reduce the wait time for emergency messages, terminals will immediately retransmit emergency messages. Terminals transmitting e-mail requests will wait between one and eight time slots before retransmitting.

When the payload receives a message on the request channel, it determines whether the message is an emergency message or an e-mail request. If the payload receives an emergency message, it repeats the message in transmissions to the terminal and the gateway. When the terminal receives the repeated emergency message from the payload and determines there are no errors, the protocol is complete. If the terminal detects errors in the emergency message from the payload, it begins the protocol again until the emergency message transmission is successful.

If the payload receives an e-mail request, it sends a slot assignment (SA) to the terminal. The slot assignment includes the assignment channel and time slots in which the

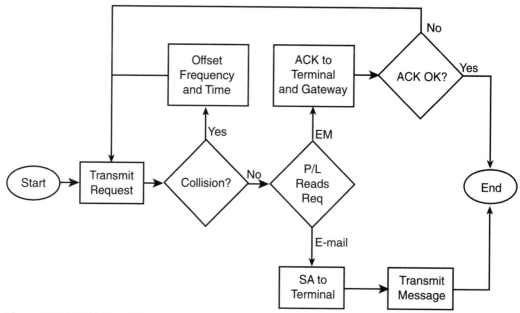

Figure 13.9 DAMA Flow Diagram

terminal can transmit its e-mail message. After the terminal transmits its e-mail message, the protocol is complete.

An example of a DAMA protocol similar to Telestar's shown in frequency and time is shown in Figure 13.10. Examining only successful transmissions, terminals 1 through 7 make e-mail requests and receive slot assignments. The time slots containing the letters A, B, and C reflect previous slot assignments.

The next step is to determine the structure of the request and assignment channels. The structure of the two channel types will determine the distribution of request channels to frequency channels.

Request Channel Structure: Making the length of emergency and request messages the same creates a scenario suited for Aloha random access. The emergency and request messages will be short, infrequent, random transmissions of the same length. The longer e-mail messages will be on different channels and will not, therefore, cause an unacceptable probability of collision with the shorter emergency and request messages.

There are two types of Aloha random access: pure Aloha and slotted Aloha. Pure Aloha, as explained previously, allows users to transmit messages at any time. In contrast, slotted Aloha requires terminals to transmit in time slots. In slotted Aloha, when messages collide, they overlap completely instead of partially. The result is a doubling in the utilization rate compared to pure Aloha.

The utilization of the request channel is the percent of time the payload receives messages without collision. For pure Aloha, the utilization, G, depends on the number of successful requests, λ, and the packet length, T, which Pratt [2] defines as:

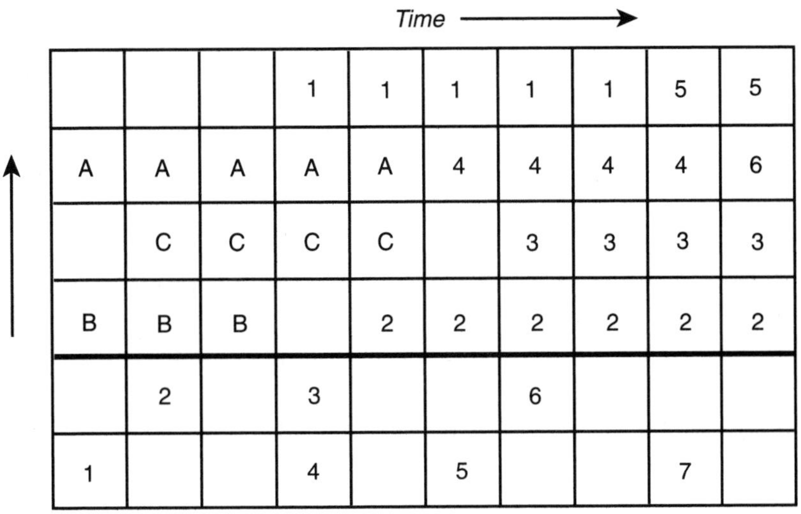

Time ⟶

			1	1	1	1	1	5	5
A	A	A	A	A	4	4	4	4	6
	C	C	C	C		3	3	3	3
B	B	B		2	2	2	2	2	2
	2		3			6			
1			4		5			7	

Figure 13.10 Example DAMA Protocol

$$G = \lambda T \qquad\qquad 13.3$$

Given a Poisson distribution, depends on the number of attempted requests, which Pratt defines as:

$$\lambda = \lambda e^{-2\lambda T} \qquad\qquad 13.4$$

The average number of times a terminal will have to transmit a packet, N, is [2]:

$$N = e^{2\lambda' T} \qquad\qquad 13.5$$

Combining the three equations yields the utilization for pure Aloha:

$$G = \tfrac{1}{2N} \log_e (N) \qquad\qquad 13.6$$

Because colliding packets in slotted Aloha collide entirely, the utilization for slotted Aloha is twice that of pure Aloha and is defined as:

$$G = \tfrac{1}{N} \log_e (N) \qquad\qquad 13.7$$

In order to double utilization of the request channels, Telestar will use slotted Aloha as the multiple-access scheme for the request channels. Figure 13.11 compares the utilization curves for Aloha and slotted Aloha, given a packet length of T = 0.065 seconds (for reasons explained later).

From Figure 13.11 and calculations for pure and slotted Aloha, the maximum utilization is 18.4% for pure Aloha with 8 attempted transmissions per second and 36.8% for slotted Aloha with 16 transmissions per second. To take advantage of the increased utilization of slotted Aloha, Telestar will use the clock already provided by the global positioning system (GPS) system to synchronize terminals and payload.

Assignment Channel Structure: The assignment channel is a TDMA scheme with terminals taking turns transmitting according to their slot assignments. A major of advantage of TDMA is its high throughput compared to FDMA and CDMA, as shown in Figure13.8. Telestar's slotted Aloha protocol for the request channels allows for implementation of TDMA for the assignment channels, since both require the use of time slots to transmit packets. The request and assignment channels will be synchronous, as shown in the example in Figure 13.10 on page 223.

Number of Request Channels: There is a trade-off between the number of request channels and the average length of e-mail messages. Having more request channels allows for more users on the system but provides for fewer resources to handle e-mail messages.

The number of request channels required is based on the ratio, x, of the e-mail message length to the request message length. The ratio of the number of assignment channels, AC, to the number of request channels, RQ, is:

$$AC = 0.37x\text{RQ} \qquad\qquad 13.8$$

where 0.37 is the maximum utilization of the request channels. Given that there are twenty frequency channels, so that AC + RQ = 20, solving for AC yields:

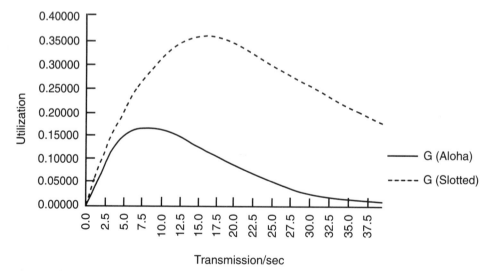

Figure 13.11 Analysis of Aloha

$$AC = {}^{7.4x}/_{(1+0.37x)}$$ 13.9

The ratio x is:

$$x = 8L_e/L_r$$ 13.10

where 8 is the number of bits per character, L_e is the number of characters in an e-mail message, and L_r is the number of bits in a request message (75 bits as explained later). Solving for RQ for different message lengths yields the curve shown in Figure 13.12.

The more request channels available on Telestar, the more requests the system can handle. It is therefore desirable to maximize the number of request channels in order to provide service to the maximum number of users. As shown in Figure 13.12, however, there is a tradeoff from the number of request channels to the average length of e-mail messages. Choosing four request channels allows for an average e-mail message length of about 101 characters. This should be sufficient, given that some messages will be very short and emergency messages will not require any resources on the assignment channels.

Daily Request Capacity: Given four request channels, we can determine the number of original requests the system can handle in one day. This daily request capacity, C, is based on the availability of coverage and the number of original requests, λ, the system will receive per second. The daily request capacity is defined as:

$$C = 86,400(0.39)RQ\lambda$$ 13.11

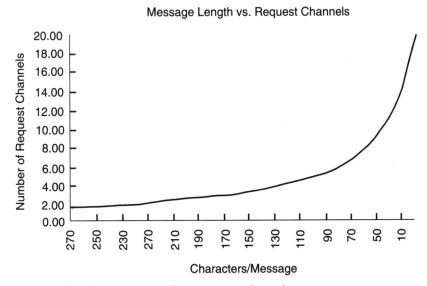

Figure 13.12. Message Length vs. Request Channels

where 86,400 is the number of seconds in a day and 0.39 is the availability of coverage as a percentage. Figure 13.13 shows the daily capacity of original requests for λ at maximum efficiency and a varying number of request channels.

From Figure 13.13, Telestar's four request channels can handle almost 800,000 original requests per day.

The overall throughput of the subscriber uplink channels is a combination of the throughput of the request and assignment channels. Using a throughput of 90 perecent from Figure 13.8 for the 16 TDMA assignment channels and 37 percent for the 4 request channels, the overall throughput, μ, is:

$$\mu = 0.9(0.75) + 0.37(0.25) = 0.77 \qquad\qquad 13.12$$

Assuming maximum usage, Telestar's overall throughput for the subscriber uplink is approximately 77 percent.

Subscriber Downlink Structure: The subscriber downlink for Telestar will carry Aloha protocol acknowledgments and slot assignments as well as subscriber messages from the payload to terminals. As stated previously, Telestar has eight subscriber downlink channels. In order for the terminal to receive a message from the payload, it must either listen to all eight channels simultaneously or listen to the one channel on which the payload will transmit its message. Although the latter requires some overhead to coordinate between terminal and payload, it saves receiver power because the terminal continuously demodulates only one frequency channel compared to eight. To save terminal battery power, Telestar will implement a simple scheme to coordinate transmissions from the payload to the terminal.

Similar to the subscriber uplink, the downlink will use TDMA in addition to the eight FDM downlink channels. During its assembly, each terminal will

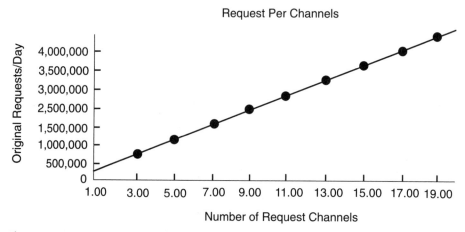

Figure 13.13 Request Capacity for a Given RQ

receive a permanent identification and a single demodulator for only one of the eight channels. Terminals permanently assigned to the same channel will be randomly distributed across the subscriber area. The payload will store in memory each terminal's identification number and which downlink channel it demodulates. When the payload needs to transmit a message to a terminal, it reads the terminal identification and transmits on the appropriate downlink channel. Terminals listen continuously on their assigned channels. When a terminal detects packets containing its terminal identification in the address, it processes the messages and stores them for retrieval by the subscriber.

The throughput for the TDMA subscriber downlink channels depends on the traffic. When fully loaded, throughput could be over 90 percent as shown in Figure 13.8.

Gateway Uplink and Downlink: The data rate for the uplink and downlink between the payload and the gateways is 56 kbps. Telestar's DAMA multiple-access scheme must allow for sufficient capacity on the gateway channels for command and control messages. The maximum capacity on the gateway links required for the subscriber uplink and downlink, given expected throughput rates, will be approximately 37 kbps. The remaining capacity of 19 kbps is available for command and control traffic, the structure of which the ground station design team has defined.

13.4.3 Message Structure

The packet length for the subscriber uplink and downlink depends on the data rate and coding rate. From the system architecture, the uplink data rate is 2400 bits per second (bps) and the downlink data rate is 4800 bps. Telestar will use a coding rate of one-half, as explained later.

Uplink Message Description: Request messages must contain information to inform the payload which terminal sent the message, the message type, and appropriate information for each type of message. Emergency messages must contain location information. E-mail request messages will contain the amount of resources required and a message priority level. The e-mail message on assignment channels will contain only data bits, which include the e-mail destination address and the message itself.

The bit field for request messages is listed in Table 13.2.

As shown in Table 13.2, 22 terminal identification bits allow for over 4.1 million subscribers and accommodate growth in the system. The 22 bits for both latitude and longitude provide location accuracy within 30 meters. The four priority levels allow subscribers to assign priority to their e-mail messages through the Internet.

Because the uplink message bit field contains 75 bits of information and the coding rate is one-half, each uplink packet will contain 150 bits. Given 150 bits per packet and an uplink data rate of 2400 bps, each packet will be 0.0625 seconds long. This packet length allows for 16 packets per second. Assignment channel packets will also contain 75 information bits and 75 coded bits.

Downlink Message Description: Subscriber downlink packets must allow for the two types of messages from the payload to terminals: acknowledgments and subscriber

Table 13.2 Uplink Request Message Bit Field

Field Name	# of Bits
Terminal ID (4.1 million subscribers)	22
Message type (EM or e-mail)	1
E-mail message length	6
Latitude	22
Longitude	22
Priority	2
TOTAL	75

messages. Each packet must contain the terminal identification and message type. The bit field for both is listed in Table 13.3.

If the message is an acknowledgment, the payload assigns the terminal an assignment frequency channel on which to transmit and the number of time slots to wait before transmitting the e-mail message. The number of slots field will be the same as those requested in the uplink e-mail request.

Having defined the modulation modes, multiple-access scheme and message structure, we must next design an error detection and correction method to ensure an acceptable bit error rate.

Table 13.3 Downlink Message Bit Field

Field Name	# of Bits
Terminal ID	22
Message type (ACK or subscriber message)	1
If ACK (slot assignment)	
Frequency channel	4
Start time slot	8
Number of slots	6
Spare	109
Subtotal	**150**
If subscriber message	
Message data	127
Subtotal	**150**

13.4.4 Error detection and correction

The transmission of information over a satellite communications system always results in some degradation in the quality of information. Unlike analog systems where it is not possible to remove noise from the signals, in digital systems errors can be detected by adding redundant bits to the data stream to locate errors. The degradation is measured in terms of bit error rate (BER). BER is the probability that a bit sent over the link will be received incorrectly and is expressed as a single number.

Design considerations: The design considerations for Telestar I satellite system error detection and correction scheme are envisioned as follows:

1. Minimize errors in transmission by optimum choice of error detection and correction scheme. A target specification for bit error rate of 10^{-6}, which is the standard rate for similar satellite systems, has been defined for Telestar.
2. Minimize power at user terminal, because a small hand-held user terminal has been recommended for the system..
3. Adding of redundant bits to the data, also called coding, results in lowering the error rate but it also causes reduction of throughput. Throughput to be optimized for the defined error rate.
4. Minimize system complexity and transmission delays. Payload processing load to be optimized for performance.

Decoding and error correction: The decoder uses the redundancy at the encoder in order to detect and correct errors. In this respect, two techniques can be distinguished and can be used: forward error correction (FEC) and automatic repeat request (ARQ).

13.4.5 Forward error correction versus automatic repeat request

The two above-mentioned techniques differ in their application. Systems that can locate and correct errors use forward error correction. Techniques that require retransmission of corrupted data blocks are called automatic repeat request. FEC is the decoding mode recommended for Telestar I for the following reasons:

- **Throughput.** Throughput is defined as the ratio of the number of bits of data transmitted to the maximum number of data bits that can be sent over the system. In ARQ schemes, every time an error is detected, the whole block has to be retransmitted. This reduces the throughput to a great extent. It varies from *0.03* for stop-and-wait ARQ to *0.84* for Selective-Repeat ARQ. In the case of FEC, the throughput remains higher than ARQ because no retransmission is involved.
- **Buffer requirements.** In ARQ, a certain amount of on-board processing is needed to detect errors so that retransmission can be ordered. There is also a requirement of storage in terms of buffer memory in the satellite link to hold data. This increases the complexity at the processor end, which is not desirable.
- **Delays.** In order to be able to order and receive the retransmission, ARQ requires a return path. This increases traffic and leads to delays.

- **Real-time satellite transmission.** ARQ techniques are easy to apply to terrestrial links where data is not transmitted real-time. In the case of FEC means are available to both detect and correct errors and is therefore suitable for systems such as Telestar where real-time transmission is needed.

Bandwidth-power interchange: The payoff for using coding techniques that reduce Eb/No required for a given bit error probability is an increase in allowable data rate and/or a decrease in necessary received C/No. If the code rate is defined as the ratio of the information bits to the total number of bits, where *n* is the number of data bits and *r* is the number of redundant bits, the code rate would be n/(n+r). The effect of coding is illustrated in Figure 13.14, where the data rate has been shown as a function of C/No for a constant bit error probability. The figure compares the effect with and without coding and shows that reducing the data rate widens the bandwidth and enables power to be economized by reducing the carrier-to-noise ratio.

Encoding and decoding schemes: In designing a communication system to operate at a specified data rate, the improvement in efficiency to be realized using coding must be weighed against the relative costs. Potential alternatives include increasing the transmitted power, increasing the transmitting antenna gain, and/or the receiving antenna area and accepting a higher probability of bit error. For Telestar, minimum received power is required because of the limitation of the small hand-held receiver. The incremental cost per decibel increase in C/No would be greater than the cost of reducing the Eb/No through coding. The two coding schemes available are block and convolutional encoding. Block coding is of two types: linear and cyclic. Linear block encoding is inefficient for multiple errors because the implementation becomes complex as the number of errors increases.

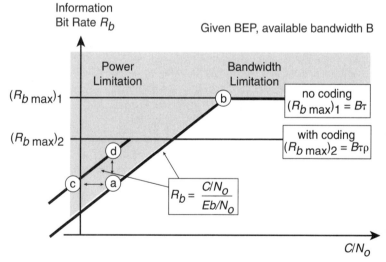

Figure 13.14 Data Rate as a Function of C/No at Constant Error Probability

Cyclic codes, such as BCH, are easier to implement because the codes can be generated by using shift registers, but only a fixed number of errors can be detected for a certain block length. Another limitation with block coding is that only a limited number of errors can be corrected (e.g., for linear block coding only half the numbers of errors detected can be corrected). The error detection and correction scheme proposed for Telestar is convolutional encoding with Viterbi decoding.

13.4.6 Convolutional encoding

Convolutional codes are generated by a tapped shift register and two or more modulo-2 adders wired in a feedback network. The output is the convolution of the incoming bit stream and the impulse response of the shift register and its feedback network. As each incoming bit propagates through the shift register, it influences several outgoing bits, spreading the information content of each data bit among several adjacent bits. An error in any one output bit can be overcome at the receiver without any information being lost.

In the encoder, information bits are shifted into register k bits at a time. After each k bits the output of the modulo-2 adders are sampled sequentially, yielding the code symbols. Since v code symbols are generated for each set of k information bits, the code rate is k/v information bits per code symbol, where $k<v$. The constraint length of the code is K, since that is the number of k bit shifts over which a single information bit can influence the encoder output. The state of the convolutional encoder is defined by the shift register contents that will remain after the next input bits are clocked in.

The encoder proposed for Telestar is a rate one-half convolutional encoder (Figure 13.15) of constraint length 5. The values for the different functions would be $K = 5$, $k = 1$, and $v = 2$. The first four coder stages specify the state of the encoder; thus, there will be $2^{k(K-1)}$ or 16 states. The state transitions for the Telestar encoder are given in Table 13.9. Thus, an input sequence of 10110… would generate an output sequence of 1110101100., assuming that the encoder was in state 0000 at time 0. The code words, or sequence of code symbols, generated by the encoder for various input information bit sequences can be shown as a code "trellis," which is actually a form of state diagram representation. The lines or "branches" joining states indicate state transitions due to input of single information bits.

Viterbi decoding: A decoder keeps track of the encoder's state transitions and reconstructs the input bit stream. Transmission errors are detected because they correspond to a sequence of transitions that could not have been transmitted. When an error is detected, the decoder begins to construct and keep track of all the possible tracks, or sequence of state transitions, that the encoder may be transmitting. At some point, which depends on its speed and memory, the decoder selects the most probable track and puts out the input bit sequence corresponding to that track. Unfortunately, the number of paths for an L bit sequence is 2^L. This brute force decoding quickly becomes impractical as L increases. The Viterbi decoding algorithm greately reduces the effort required for maximum likelihood decoding by making use of the special structure of the code trellis.

The trellis assumes a fixed periodic structure after a depth of 5, in general, K, is reached. These paths are said to have diverged at state 0000, depth 0 and remerged at state 0000, depth 5. Paths remerge after k(K-1) consecutive identical information bits. A Viterbi

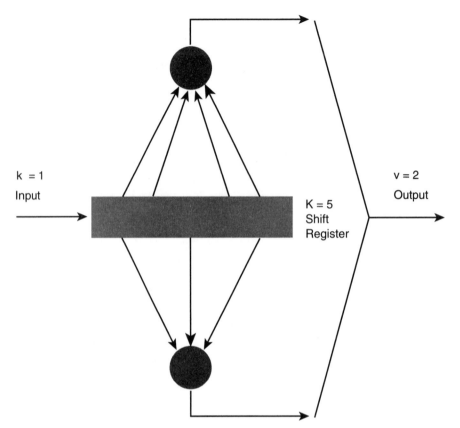

Figure13.15 Rate One-half Convolutional Encoder with Constraint Length of 5

decoder calculates the likelihood of each of the 2^k paths entering a given state and eliminates from further consideration all but the most likely path that leads to that state. This is done for each of the $2^{k(K-1)}$ states at a given trellis depth; after each decoding operation, only one path remains leading to that state. The decoder then proceeds one level deeper into the trellis and repeats the operation. The fact that all the paths tend to have a common stem also suggests that if the decoder stores enough of the past information bit history of each of the $2^{k(K-1)}$ paths, then the oldest bits on all the paths will be identical. If a fixed amount of path history is provided, the decoder can output the oldest bit on an arbitrary path each time it steps one level deeper into the trellis. The amount of path storage required, u, is equal to the number of states multiplied by the length of the information bit path history per state h, $u = h2^{k(K-1)}$.

Since the path memory represents a significant portion of the total cost of a Viterbi decoder, it is desirable to minimize the required path history length h. It has been demonstrated theoretically and through simulation that a value of h of four or five times the code constraint length is sufficient for negligible degradation from optimum decoder performance.

Table 13.9 State Transitions in Telestar Convolution Encoder

Initial State	Input	Register after input	Output	Ending State
a(0000)	0	00000	00	a(0000)
	1	10000	11	I(1000)
b(0001)	0	00001	11	a(0000)
	1	10001	00	I(1000)
c(0010)	0	00010	10	b(0001)
	1	10010	01	j(1001)
d(0011)	o	00011	01	b(0001)
	1	10011	10	j(1001)
e(0100)	0	00100	01	c(0010)
	1	10100	10	k(1010)
f(0101)	0	00101	10	c(0010)
	1	10101	01	k(1010)
g(0110)	0	00110	11	d(0011)
	1	10110	00	I(1011)
h(0111)	0	00111	00	d(0011)
	1	10111	11	I(1011)
I(1000)	0	01000	10	e(0100)
	1	11000	01	m(1100)
j(100)	0	01001	01	e(0100)
	1	11001	10	m(1100)
k(1010)	0	01010	00	f(0101)
	1	11010	11	n(1101)
I(1011)	0	01011	11	f(0101)
	1	11011	00	n(1101)
m(1100)	0	01100	11	g(0110)
	1	11100	00	o(1110)
n(1101)	0	01101	00	g(0110)
	1	11101	11	o(1110)
o(1110)	0	01110	01	h(0111)
	1	11110	10	p(1111)
p(1111)	0	01111	10	h(0111)
	1	11111	01	p(1111)

13.4.7 Performance Parameters

Quantization levels: The binary symbols output by the encoder are used to modulate an RF carrier sinusoid. For BPSK, which is the modulation scheme proposed by the signal processing team, each code symbol results in the transmission of a pulse of the carrier at either of two 180 degree separated phases. To facilitate digital processing by the decoder, this continuous signal r_j is greater than 0 and 1 output otherwise. Here, the received data are represented by only one bit per code symbol. When coding is used, hard decision of the received data usually entails a loss of 2 dB in Eb/N_o compared with infinitely fine quantization. Much of this loss can be recouped by quantizing r_j to four or eight levels instead of merely two. Adding additional levels of quantization necessitates at 2- or 3-bit representation for each r_j. Figure 13.16 compares the performance for two-, four-, and eight-level quantization. Eight-level quantization has been proposed for Telestar.

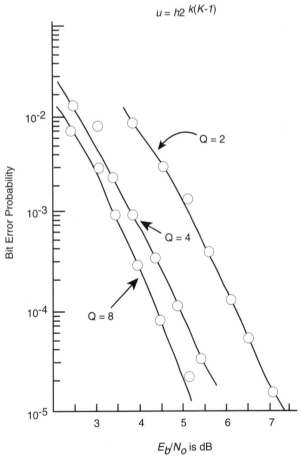

Figure 13.16 Performance Comparison of Viterbi Decoding, Using rate 1/2, K=5 code with 2-, 4-, and 8-level Quantization

Constraint length: One of the factors to be considered in design is to achieve a low Eb/No for a specified BER. An improvement in Eb/No of 0.4 to 0.5 dB can be obtained by increasing the constraint length or span of the encoder by one. Figure 13.17 shows the performance curves for 8-level quantization for different constraint lengths. However, increasing the constraint level by 1 more than doubles the decoding complexity. In recent years, this problem has been greatly reduced by hardware technological advancement. For Telestar a constraint level of 5 has been proposed.

Code rate: Code rates less than one-half buy improved performance at the expense of increased bandwidth expansion and more difficult symbol tracking due to decreased symbol energy-to-noise ratios. Rates above one-half conserve bandwidth but are less efficient in energy. Rate one-half encoding is economical and simple to implement and has been recommended for Telestar.

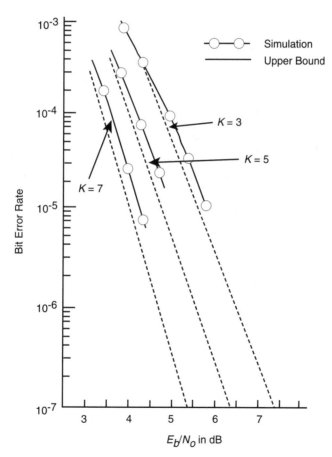

Figure13.17 Bit Error Rate versus Eb/No for Rate 1/2 Viterbi Decoding. 8-level Quantized Simulations with K = 3, 5, 7

Decoding gain: The decoding gain is defined as the difference in Eb/No, at a given value of bit error rate, between transmissions with and without encoding. The advantage to the link budget provided by the decoding gain is paid for by an increase in the bandwidth used on the space link. Telestar is power limited rather than bandwidth limited, and thus decoding gain can be used to improve performance.

1. Telestar uplink

   (a) In the absence of coding (p=1). The transmitted bit rate Rb = 2.4 kbps. The modulation used is BPSK, so the spectral efficiency would be I = 1.0 bps/Hz. The bandwidth used is B = 2.4 kbps / 1.0 bps/Hz = 2.4 Khz. The required value of C/No is:

   (C/No) = (Eb/No) * Rb

   To ensure that BER = 10^{-6}, (Eb/No) = 10.5 dB if a demodulator implementation degradation of 1.5 dB is considered.

   (Eb/No) = 10.5 + 1.5 = 12 dB

   (C/No) = 12 + 10 log 2400

   = 45.8 dB

   (b) With coding at rate one-half (p=1/2). The binary transmission rate is Rb/p = 1.2 * 2 = 2.4 kbps. The bandwidth used is B = 2.4/1.0 = 2.4 kHz. The (Eb/No) achieved for BER = 10^{-6} with Viterbi decoding is 5.7 dB.

   (Eb/No) = 5.7 + 1.5 = 7.2 dB

   (C/No) = (Eb/No) * Rb

   = 7.2 + 10 log 2400 = 41.00 dB

   (c) Decoding gain for uplink = 45.80 - 41.00 = 4.80 dB

2. Telestar downlink

   (a) In the absence of coding (p=1). The tranmitted bit rate Rb = 4.8 Kbps. The bandwidth used is B = 4.8 kbps/1.0 bps Hz = 4.8 kHz. The required value of C/No is:

   (C/No) = (Eb/No) * Rb

   To ensure that BER = 10^{-6}, (Eb/No) = 10.5 dB if a demodulator implementation degradation of 1.5 dB is considered.

   (Eb/No) = 10.5 + 1.5 = 12 dB

$$(C/No) = 12 + 10 \log 4800$$
$$= 48.81 \text{ dB}$$

(b) With coding at rate one-half (p = 1/2). The binary transmission rate is Rb/p = 2.4 * 2 = 4.8 kbps. The bandwidth used is B = 4.8/1.0 = 4.8 kHz. The (Eb/No) achieved for BER = 10^{-6} with Viterbi decoding is 5.7 dB.

$$(Eb/No) = 5.7 + 1.5 = 7.2 \text{ dB}$$
$$(C/No) = (Eb/No) * Rb$$
$$= 7.2 + 10 \log 4800 = 44.01 \text{ dB}$$

(c) Decoding gain for downlink = 48.81 - 44.01 = 4.80 dB

13.4.8 Telestar scheme for error detection and correction

Error detection and correction are important functions of any satellite systems. For the Telestar system, which will be marketed primarily as an emergency messaging system, it is important to ensure accuracy of data and provide minimum delays. Forward error correction is the technique proposed for Telestar, as it provides better throughput and avoids delays due to retransmission as compared to automatic repeat request. Convolutional encoding and Viterbi decoding would perform better than linear and cyclic block coding, considering the unique needs of this system, and can be implemented using digital hardware. Rate one-half encoding and a constraint length of five for the convolution encoder allows an improvement of 4.2 dB in Eb/No and a decoding gain of 4.8 dB at the receiver for a bit error rate of 10^{-6}.

13.4.9 Summary

The Telestar Satellite System is designed as a commercially viable Low Earth Orbit (LEO) system providing critical emergency aid and routine electronic messages. Based on marketing and FCC filings, the Telestar system was designed as an economical, efficient, and low-risk means to provide mobile personal messaging using the latest technical knowledge. The signal processing design of the Telestar payload is the heart of the system. Designed with BPSK modulation, demand assignment, foward-error correction, convolution encoding, and Viterbi decoding, the Telestar signal processing design will ensure that the most customers are served in the shortest time with the utmost accuracy. As a state-of-the-art LEO system, Telestar is the cutting edge of the emerging personal communications industry.

Appendix A to Chapter 13
Satellite System Design Methodology

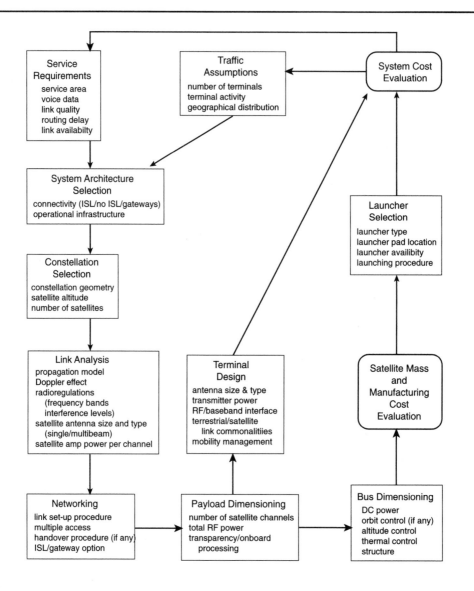

Appendix B to Chapter 13

Aloha and Slotted Aloha Calculations

lamda'	lamda	lambda (Slotted)	N (Aloha)	G (Aloha)	N (Slotted)	G (Slotted)
0.0	0.000	0.0000	1.00000	0.00000	1.00000	0.00000
0.5	0.470	0.4846	1.06449	0.02936	1.03174	0.03029
1.0	0.882	0.9394	1.13315	0.05516	1.06449	0.05871
1.5	1.244	1.3658	1.20623	0.07772	1.09829	0.08536
2.0	1.558	1.7650	1.28403	0.09735	1.13315	0.11031
2.5	1.829	2.1384	1.36684	0.11431	1.16912	0.13365
3.0	2.062	2.4871	1.45499	0.12887	1.20623	0.15544
3.5	2.260	2.8123	1.54883	0.14124	1.24452	0.17577
4.0	2.426	3.1152	1.64872	0.15163	1.28403	0.19470
4.5	2.564	3.3968	1.75505	0.16025	1.32478	0.21230
5.0	2.676	3.6581	1.86825	0.16727	1.36684	0.22863
5.5	2.766	3.9001	1.98874	0.17285	1.41023	0.24376
6.0	2.834	4.1237	2.11700	0.17714	1.45499	0.25773
6.5	2.884	4.3299	2.25353	0.18027	1.50118	0.27062
7.0	2.918	4.5195	2.39888	0.18238	1.54883	0.28247
7.5	2.937	4.6934	2.55359	0.18357	1.59800	0.29334
8.0	2.943	4.8522	2.71828	0.18394	1.64872	0.30327
8.5	2.938	4.9969	2.89360	0.18360	1.70106	0.31231
9.0	2.922	5.1280	3.08022	0.18262	1.75505	0.32050
9.5	2.897	5.2464	3.27887	0.18108	1.81077	0.32790
10.0	2.865	5.3526	3.49034	0.17907	1.86825	0.33454
10.5	2.826	5.4473	3.71545	0.17663	1.92755	0.34046

lamda'	lamda	lambda (Slotted)	N (Aloha)	G (Aloha)	N (Slotted)	G (Slotted)
11.0	2.781	5.5311	3.95508	0.17383	1.98874	0.34570
11.5	2.731	5.6047	4.21016	0.17072	2.05187	0.35029
12.0	2.678	5.6684	4.48169	0.16735	2.11700	0.35427
12.5	2.620	5.7229	4.77073	0.16376	2.18420	0.35768
13.0	2.560	5.7687	5.07842	0.15999	2.25353	0.36054
13.5	2.497	5.8063	5.40595	0.15608	2.32507	0.36289
14.0	2.433	5.8361	5.75460	0.15205	2.39888	0.36475
14.5	2.367	5.8585	6.12574	0.14794	2.47502	0.36616
15.0	2.300	5.8741	6.52082	0.14377	2.55359	0.36713
15.5	2.233	5.8831	6.94138	0.13956	2.63465	0.36770
16.0	2.165	5.8861	7.38906	0.13534	2.71828	0.36788
16.5	2.098	5.8833	7.86561	0.13111	2.80457	0.36770
17.0	2.030	5.8750	8.37290	0.12690	2.89360	0.36719
17.5	1.963	5.8618	8.91290	0.12272	2.98545	0.36636
18.0	1.897	5.8437	9.48774	0.11857	3.08022	0.36523
18.5	1.832	5.8213	10.09964	0.11448	3.17799	0.36383
19.0	1.767	5.7947	10.75101	0.11045	3.27887	0.36217
19.5	1.704	5.7642	11.44439	0.10649	3.38296	0.36026
20.0	1.642	5.7301	12.18249	0.10261	3.49034	0.35813
20.5	1.581	5.6926	12.96820	0.09880	3.60114	0.35579
21.0	1.521	5.6521	13.80457	0.09508	3.71545	0.35325
21.5	1.463	5.6086	14.69489	0.09144	3.83339	0.35054
22.0	1.406	5.5625	15.64263	0.08790	3.95508	0.34765

REFERENCES

[1] Ha, Tri T. *Digital Satellite Communications*. New York: Macmillan, 1986.

[2] Pratt, Timothy, and Charles W. Bostian. *Satellite Communications Systems*. (New York: John Wiley and Sons, 1986.)

[3] Maral, G., and M. Bousquet. *Satellite Communications Systems*. (Chichester, England: John Wiley, 1984.)

[4] Couch, Leon W. *Digital and Analog Communications Systems*. (New York: Macmillan, 1990.)

Chapter 14

Mobile Cellular Communications Code Division Multiple Access (CDMA)Application and Implementation

With the increasing demand for mobile cellular communications on the current analog Advanced Mobile Phone System (AMPS), the cellular communications industry is investigating means of providing better quality and greater quantities of cellular service. AMPS presently uses a Frequency Division Multiple Access (FDMA) technique by dividing the available cellular communications spectrum (824–849 and 869–894 MHz) into 30 kHz-wide channels. Since the frequency allocation for the current cellular services are fixed, alternatives to the present FDMA system are required in order to meet the expanding number of cellular subscribers. Alternatives being considered are overlaying a Time Division Multiple Access (TDMA) scheme on top of the existing FDMA system or going entirely to a Time Division Multiple Access (TDMA) approach. However, these alternatives suffer from many of the same limitations as the current FDMA system, such as frequency reuse restrictions, multipath fading, and cochannel interference. Another alternative receiving much attention is the digital code division multiple access (CDMA) technique recently standardized by the Telecommunications Industry Association for the North American cellular telephone system as IS–95. This technique offers capacity and quality advantages over the present system, but not without its own set of constraints, namely, cost and implementation. This chapter describes the cellular CDMA approach as compared to FDMA, discusses its advantages over other techniques, and highlights the disadvantages and challenges of implementing a CDMA system. In particular, the Globalstar CDMA system is discussed to show the application of a Low Earth Orbit (LEO) systems.

14.0 MULTIPLE ACCESS COMPARISON AND DESCRIPTION

The purpose of multiple access is to divide up the available frequency-time space among n users while minimizing the interference between them. Assuming a transmission interval of T and a total available bandwidth of B, the maximum number of n users is expressed as follows:

$$n = 2BT \qquad\qquad 14.1$$

For FDMA, each user receives an individual frequency channel with transmission time intervals, T_i, given by

$$T_i = T \qquad\qquad 14.2$$

which is to say that user transmissions can be continuous with respect to the total available time interval, T. On the other hand, the individual user bandwidth interval, B_i, is given by

$$B_i = B/2 = 1/2T$$

which is to say that individual user transmissions occupy only a portion, or channel, of the total available bandwidth, B. FDMA is graphically represented in Figure 14.1.

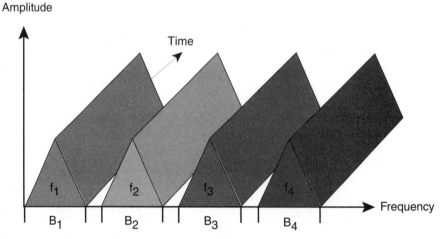

Figure 14.1 FDMA

For TDMA, the users share the same frequencies but transmit at different time intervals such that the individual user bandwidth interval, B_i, is given by

$$B_i = B \tag{14.3}$$

which is to say that individual users may transmit at any frequency within the band, and individual user transmission time intervals, T_i, is given by

$$T_i = T/2 = 1/2B \tag{14.4}$$

which is to say that the user transmissions are scheduled or periodic. TDMA is graphically represented in Figure 14.2.

With CDMA, on the other hand, users share the entire available bandwidth and transmission interval such that

$$T_i = T \tag{14.5}$$
$$B_i = B$$

Instead of using frequency and/or time diversity, CDMA achieves multiple access by assigning each user a unique, orthogonal transmission code. Thus, multiple users can transmit simultaneously using the same frequency band, and the signal can be decoded provided the receiver code allows the transmitted signal to be extracted from the overall frequency-time space. CDMA is graphically represented in Figure 14.3.

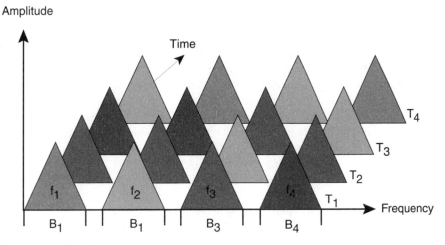

Figure 14.2 TDMA

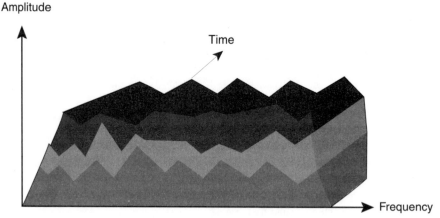

Figure 14.3 CDMA

14.1 CDMA ADVANTAGES

The advantages of CDMA in cellular applications are improved frequency reuse (which ultimately translates into improved channel capacity), reduced multipath fading, cochannel interference "elimination," graceful degradation (i.e., no set capacity limit) and soft hand-off. The following paragraphs briefly describe each of these advantages in more detail. The ultimate advantage would be to have a land based and satellite CDMA system.

14.1.1 Frequency Reuse

Because frequency reuse is required in a cellular system, much attention is paid to frequency planning. In the present FDMA system, frequency planning involves partitioning the area to be serviced into cells, forming clusters of K adjacent cells, and allocating n/K channels to each cell, where n is the total number of channels per cluster. Then, this K cell cluster is repeated as necessary to cover the desired area. The number of cells per cluster, K, is called the frequency reuse factor, which determines the number of radio frequency channels and, therefore, amount of cochannel interference in the cell.

For AMPS, the frequency reuse factor is most frequently set to 7 in order to minimize cochannel interference and obtain the desired carrier-to-interference (C/I) ratio. While a CDMA system does not have the same problem of separating frequencies in adjacent cells to reduce interference, frequency reuse factors have been calculated for CDMA systems due to the interference power of other cells (instead of assuming a frequency reuse of 1, as is often the case with CDMA analysis). Instead of cochannel interference, a CDMA system suffers degradation because of general interference from 6 ring-1 cells, 12 ring-2 cells, 18 ring-3 cells, etc., as shown in Figure 14.4.

As derived by Kim [1], K for a CDMA cellular system is estimated at 1.33. Estimating the individual cell capacity reduction due to the frequency reuse factor as

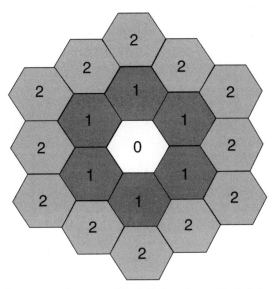

Figure 14.4 Ring 1 and 2 Cells Interfere with Cell 0.

$$(K - 1)/K \qquad\qquad 14.6$$

an FDMA cellular system with frequency reuse factor of 7 experiences an individual cell capacity reduction of

$$(7 - 1)/7 = 6/7 = 86\% \qquad\qquad 14.7$$

whereas a CDMA cellular system with frequency reuse factor of 1.33 experiences an individual cell capacity reduction of

$$(1.33 - 1)/1.33 = 0.33/1.33 = 25\%$$

Thus, it can be seen that the channel capacity reduction from frequency reuse to a CDMA cellular system is far less than that of an FDMA system. Based on the above figures and 30 channels per CDMA cell [2], Figure 14.5 compares the channel capacity of 7-cell FDMA and CDMA systems.

14.1.2 Multipath Fading

One of the issues faced in the mobile radio environment is multipath or Rayleigh fading due to reception of signals reflected from surrounding structures causing random amplitude and phase variations on the received signal. This type of fading is exhibited as short-term signal drops at the receiver and is one of the more severe cellular service quality prob-

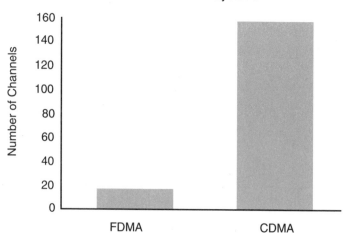

Figure 14.5 Comparison of FDMA and CDMA.
Source: Reprinted from [2].

lems. Since multipath fading may cause significant signal level drops over a bandwidth of up to 300 kHz, the impact to an FDMA cellular system is significant, since such a fade would impair 10 FDMA analog channels (i.e., 10 channels x 30 kHz/channel = 300 kHz). However, for a CDMA system with a bandwidth of 1.25 MHz, this equates to less than 25 percent of the CDMA signal. Therefore, the receiver can average out these signal fades when decoding the transmitted signal. Additionally, the CDMA system takes advantage of the multipath environment through the use of a RAKE-type receiver that weights the time-delayed signal variations in proportion to their strength and combines them optimally to overcome fading. Thus, a CDMA system is not affected by multipath fading as FDMA systems are.

14.1.3 Cochannel Interference "Elimination"

Cochannel interference is the situation where neighboring cells using the same frequencies as in a base cell interfere with effective signal reception. In the FDMA system, this is mitigated by frequency planning or by ensuring that frequencies are not used in adjacent cells (i.e., one or more cells separate the same frequencies). Nevertheless, the effect of cochannel interference is still observed in FDMA systems. In a CDMA system, since the entire bandwidth is used for each individual user, cochannel interference is not a problem. Instead, adjacent channel users increase the background interference, or noise, level. However, when the signal is received, the original signal is decoded back to the original data rate of 9600 bits-per-second (bps), while the interference signal maintains its 1.25 MHz bandwidth. This difference of 128-to-1 results in a net coding gain of 21 dB, which helps overcome problems with background interference of other CDMA users operating in the same bandwidth.

14.1.4 Graceful Degradation

In an FDMA or TDMA system, a specific capacity threshold exists. For example, 42 frequency channels exist in an FDMA system, so there can never be more than 42 users on the system at any given time. Likewise, a TDMA system has a specified number of timeslots. However, for a CDMA system, the capacity limit is "soft." That is, more and more users can be added to the system without ever reaching a set capacity limit. Instead, the increased number of users will increase the background interference level (possibly to a point where an effective capacity limit is reached, for example, communications are no longer effective due to the error rate). The advantage of this soft capacity limit is increased flexibility—flexibility to service additional users during peak hours, if desired, instead of reaching a hard capacity limit with the idea that degraded service is better than no service.

14.1.5 Soft Handoff

In an FDMA or TDMA system, when hand-off from one cell to another occurs, the user receives a new frequency channel or timeslot assignment, respectively, in the new cell. With this type of approach, the user must stop transmitting on the old cell and begin transmitting on the "new" cell at the appropriate frequency or time. Thus, communications are broken in the old cell and then reestablished in the new cell. Between these two events, brief disruptions occur in the communications. This type of hand off is referred to as a break-before-make system. In some cases, the hand-off is not successful because a new frequency or timeslot is unavailable. In a CDMA system, however, since a unique code is used for multiple access and RAKE-type receivers are used, which allows simultaneous reception of multiple signals using the same code, a make-before-break system is achievable. Therefore, before the system must hand off from the old cell, a connection has been made with the new cell. After this new cell connection is firmly established, the old cell connection is broken. This eliminates any communications disruptions during hand-off and increases the probability of a successful hand-off.

14.2 CDMA DISADVANTAGES AND CHALLENGES

One disadvantage of CDMA cellular communications is cost, and one challenge is CDMA implementation into the current cellular market. The following paragraphs briefly describe these issues in more detail.

14.2.1 Cost

FDMA (and TDMA, to a lesser extent) cellular systems have been produced and operated for several years. As a result, the technology, while constantly improving, is well understood, easy to manufacture and relatively affordable. However, the CDMA cellular system uses a more sophisticated, RAKE-type receiver to take advantage of multipath signals and perform soft hand-off. This type of receiver has not been produced or used to any large extent in cellular systems. As a result, initial production costs for such CDMA cellular receivers will be significantly more than for the current mobile cellular system.

In addition, the transition to a CDMA cellular system will involve reducing the FDMA cellular service that is currently in high demand and is widely proliferated. Since consumers will be unable to immediately replace their current cellular system with a new, more costly CDMA-compatible system, revenues will decline during this transition period. This will cause the cellular carriers to closely examine the cost of such a transition versus the future profit potential of increased subscribers on a CDMA system. Such economic decisions may delay, or even indefinitely postpone, a transition to a CDMA system.

14.2.2 Implementation

If and when the decision is made to transition the current AMPS cellular system to the increased potential of a CDMA system, issues of timing and implementation approach must be considered. The timing issue is primarily a matter of cost versus profit potential that is briefly addressed in the previous section. However, contained in the implementation approach are engineering issues with associated performance implications. One such engineering issue is interference on the existing analog channels from new CDMA channels.

Due to the proliferation of analog cellular mobile communications, it stands to reason that any transition to a CDMA system will not occur overnight. Instead, since the 1.25-MHz band for CDMA conflicts with only 10 percent of the AMPS frequency band, it is more likely that single analog cells would be converted to CDMA over a period of time. This new 1.25 MHz CDMA cell would provide approximately 30 channels [2], replacing the 6 analog channels that existed in the analog cell previously. However, the new 1.25 MHz CDMA cell will become an interferer to the remaining analog channels in adjacent cells that coexist in the same 1.25 MHz frequency band. The question of managing a cellular transition to CDMA format is addressed extensively in [2], in an effort to quantify the size of any required analog cellular guard zone to mitigate new CDMA cell(s) interfering with existing AMPS cells. The analysis addressed two cases, single and cluster CDMA cell transition, as represented by Figures 14.6 and 14.7.

Assuming the power spectral density of the CDMA cell signal is constant over the FDMA bands that are affected, Kim [2] computes the decrease, Δ, in analog carrier-to-interference ratio, C/I_A, in ring-1 through ring-4 cells. The computations are performed assuming the CDMA mobile unit power is equal to the FDMA mobile unit power as well as when the CDMA mobile unit power is 50 percent and 10 percent of the FDMA mobile unit power. Both the single and cluster CDMA cell transition scenarios are considered and are listed in Tables 14.1 and 14.2.

Assuming that an FDMA degradation of less than 1 dB is acceptable, Kim [2] concludes that one ring of buffer cells is required in both the single and cluster CDMA cell transition scenarios. However, if the CDMA mobile unit power is on the order of 10 per-

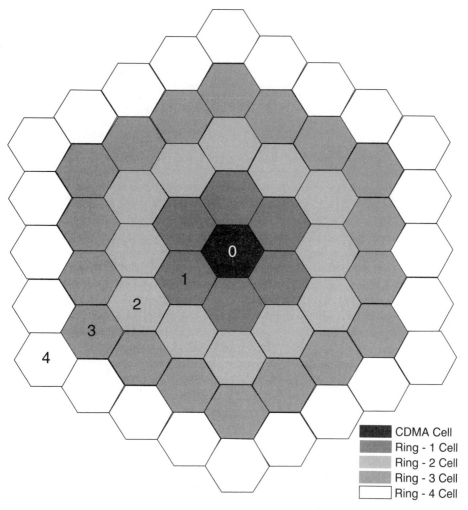

Figure 14.6 Ring-1-4 Cells for Single CDMA Cell.

cent of the FDMA mobile unit power, the ring of buffer cells may not be required at all. This was indeed the case for the 1991 CDMA field tests in which the average transmit power of the CDMA mobile units was approximately 2 percent of the peak AMPS transmit power. Prior to a decision on a cellular CDMA implementation approach, these issues of transmit power versus required buffer cell size must be resolved.

The growing demand for cellular service is causing the cellular industry to examine new approaches to satisfy this demand with a current system that has reached its capacity limits. One such alternative is a CDMA system. And while the CDMA system offers many benefits, such as improved frequency reuse, reduced multipath fading and cochannel interference, no set capacity limit, and soft hand-off capabilities, it is faced with overcoming

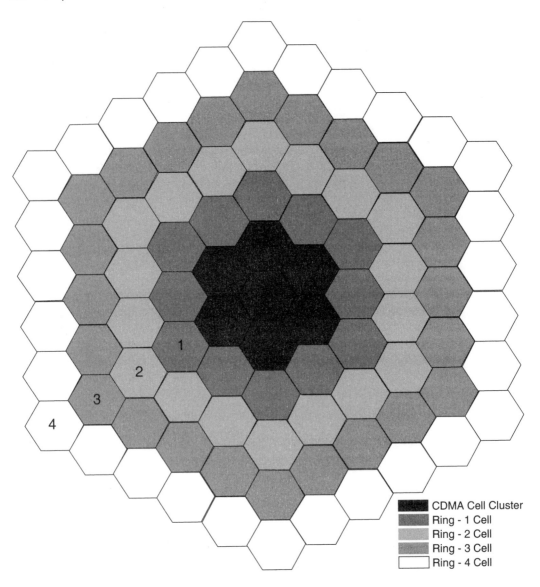

Figure 14.7 Ring-1-4 for 7-Cell CDMA Cluster.

potentially large startup costs and the inertia of a proven system with a thriving market. Only time will tell if the demand is great enough to support breaking CDMA technology into the cellular systems of tomorrow. Surely the rate at which satellite cellular and other personal communications system breakthroughs come on-line will dictate the future of CDMA in the present mobile cellular communications environment.

TABLE 14.1 AMPS Degradation in dB Due to Single CDMA Cell

	$P_C = P_A$	$P_C = 0.5 * P_A$	$P_C = 0.1 * P_A$
Ring-1 Cells	3.53	2.11	0.51
Ring-2 Cells	0.33	0.17	0.03
Ring-3 Cells	0.054	0.027	0.005
Ring-4 Cells	0.018	0.009	0.002

Table 14.2 AMPS Degradation in dB Due to 7-Cell CDMA Cluster

	$P_C = P_A$	$P_C = 0.5 * P_A$	$P_C = 0.1 * P_A$
Ring-1 Cell	5.47	3.39	0.42
Ring-2 Cell	-0.025 *	-0.39 *	-0.71 *
Ring-3 Cell	-0.59 *	-0.69 *	-0.77 *
Ring-4 Cell	0.06	0.03	0.006

* The negative values are performance improvements due to ring-2 and ring-3 cells having only five tier-1 cochannel cells instead of six as usual.

14.3 CDMA IN THE MOBILE SATELLITE SERVICE

Global cellular systems are on the verge of becoming a reality. In order to provide a viable, efficient, and cost-competitive system, many cellular designers are using spread spectrum technology in a multiple-access scheme. One such proposal is a global satellite cellular system proposed by a Loral-Qualcomm joint venture known as Globalstar. This system is designed to eventually offer voice, data and paging services worldwide. The communication technology they are proposing is CDMA, Code Division Multiple Access. CDMA is spread spectrum technology that is bandwidth efficient, resistant to interference, very secure and easily integrated into terrestrial cellular systems. This section presents the CDMA technology in general and then studies the specific reason why this technology is appropriate for a satellite cellular system.

14.3.1 The Globalstar System: An Overview

The Globalstar LEO satellite system is proposed by Loral Qualcomm Satellite Services (LQSS), a company created as a joint venture between Loral Aerospace Corp (LAC) (51%) and Qualcomm, Inc. (49%). Loral Aerospace is a large defense electronics and aerospace systems company. Qualcomm is a leading provider of CDMA digital cellular technology and equipment for terrestrial communications, as well as satellite-delivered position location and message services. The merger of the two will combine the talents of these two companies—LAC for systems design of the Globalstar system and Qualcomm for the ground segments and customer interfaces.

The Globalstar system initially calls for 32 LEOs positioned in eight polar orbiting planes, each plane containing three on-line satellites and one spare. The services to be offered are RDSS (Radio Determination Satellite Service), voice and data.

Initial operation will provide worldwide service with optimized service to the United States. It is proposed that a constellation of 48 satellites will be operational when international service becomes feasible.

In usual operation, 90 percent of satellites' traffic would be routed through a single, groundstation gateway. This system is not a "by-pass" system, and because of this, the system proposes to take advantage of the large, existing terrestrial communications infrastructure in the United States to integrate their network. No satellite crosslinks, such as in the Iridium system proposed by Motorola, will be used. The reason for this approach is obviously cost, since interoperability with existing PSTN or cellular network (Loral Cellular is developing a terrestrial cellular network as well) will reduce satellite complexity and the associated expense.

Two systems are proposed with respect to the frequency spectrum design. System A proposes use of L-band with C-band feeder links, and System B proposes both L- and S-bands with C-band feeder links. Recent action by the FCC warrants the use of System B, as it has been recommended that L-band be used for the user-to-satellite link and S-band for the satellite-to-user link. The specifics of design are included in Appendix A.

14.3.2 System Design

The Globalstar system architecture is illustrated in Figures 14.8 and 14.9. The space segment consists of the satellite launch vehicle and the constellation of microsats. The user segment includes hand-held units, as well as vehicle-mounted RDSS and voice transceivers. The ground segment includes the TT&C facilities and the gateway earth stations, which provide interface to the network control centers and local cellular gateways that open up the public-switched telephone network (PSTN) and private networks.

Projected costs for the network are around $650 million initially and $820 million for the global system. These costs are roughly broken down to $453 million for construction and launch of the initial 32 satellites, $173 million for the additional 16 satellites, $29 million for ground segment construction costs, and additional preparation, etc., costs of $174 million.

The system design is presented in five parts, each fairly brief and summarizing the components as shown in the system architecture illustration (Figure 14.8): Launch Segment, System Control, Space Vehicle Segment, Gateway Segment, and Mobile Subscriber Segment.

14.3.3 Launch Segment

The spacecrafts are designed to be launched by an expendable launch vehicle such as Delta or Arianne. Folding and stowing the antenna and solar arrays allows a stack of eight satel-

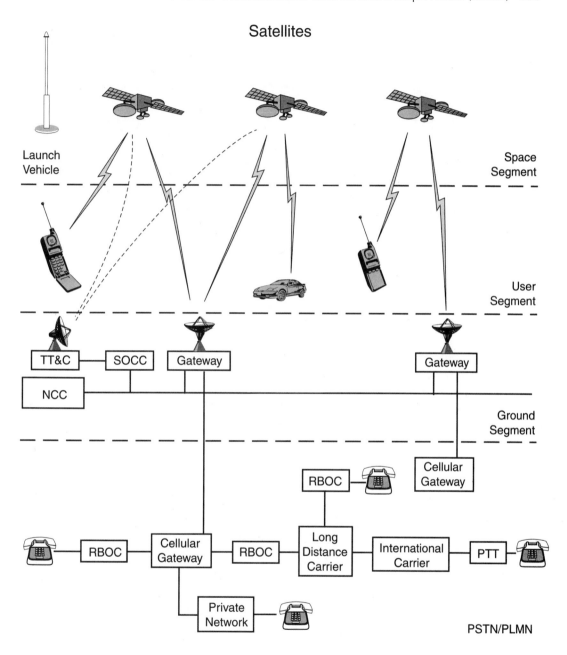

Figure 14.8 Globalstar System Architecture.

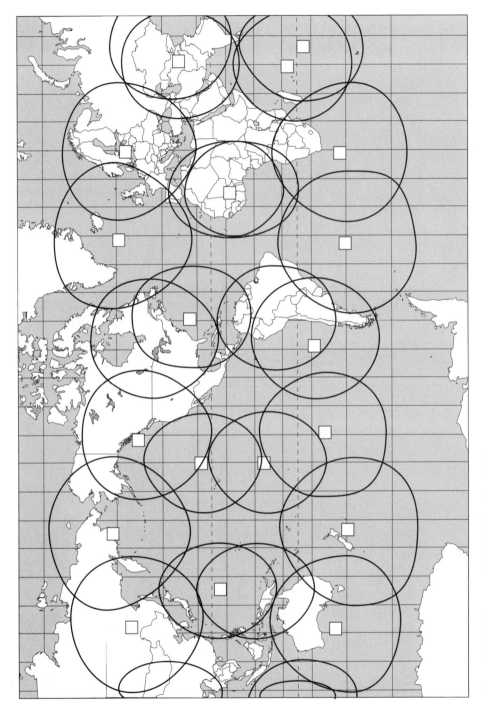

Figure 14.9 Coverage of 24 Satellite Constellation

lites to be launched in a standard McDonnell Douglas Delta or other suitable launch vehicle. The operational orbit is 1390 km and the parking orbit is 2000 km. These orbits are easily within reach of small launch vehicles and do not present some of the atmospheric drag, atomic oxygen and space debris problems that lower parking orbits (600 to 900 km) present.

The launching process is repeated until the satellite constellation is in place—32-satellites including 8 spares in four launches. This means that the entire system can be launched within one year's time. The long-term plan calls for the launching of 16 additional satellites, rounding out a 48-satellite system

The time frame for the implementation of this system is as follows:

First satellite construction	Grant + 39 months
32nd satellite construction	Grant + 57 months
8 satellites launched	Grant + 49 months
32 satellites launched	Grant + 58 months
Begin service	Grant + 60 months
16 additional satellites construction	Grant + 139 months

14.3.4 System Control

For the United States, two geographically diverse ground stations will perform the spacecraft telemetry, tracking and control. These stations, along with the Satellite Operation Control Center (SOCC), constitute the Constellation Control Center (CCC). The functions required of the CCC include orbit and de-orbit operations, status, tracking ephemeris data, and monitor and control. The TT&C facilities operating at C-band include five meter parabolic antennas to maintain positive RF link margins.

14.3.5 Space Vehicle Segment

The Globalstar satellite, shown in Figure 14.10, is equipped with a planar phased array antenna, which provides six spot beams for coverage in L-band and S-band. For the feeder links, a single beam C-band for transmitting and receiving is used. They have two-axis solar array positioning.

The major design parameters are presented in Table 14.3

The communications subsystem consists of a standard repeater design, with no on-board processing included. The antennas include separate transmit and receive C-bands for use between the satellites and gateway ground terminals for the feeder links. They are circularly polarized, with good polarization sense isolation.

There is also an L-band communications subsystem that uses six hopping elliptical spot beams. Six passive direct, radiating arrays each provide a spot beam, which provides both transmit and receive capabilities through time domain duplexing for the proposed System A. For the System B, each L-band array is receive-only.

Beam hopping and time duplexing allow for the use of the entire L-band spectrum within each spot beam.

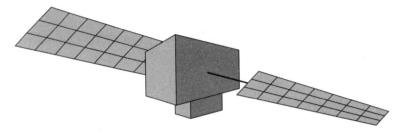

Figure 14.10 Globalstar Satellite.

TABLE 14.3 Globalstar Satellite Parameters

Stabilization	3 - Axis
Mission life	7.5 years
Station keeping	+/- 1 deg in-plane +/- 1 deg out-of-plane
Frequency bands	Refer to chapter's appendix
Nominal capacity per satellite	2,600 to 2,800 full duplex voice circuits
Total occupied bandwidth	Refer to chapter's appendix
Polarization	Circular

Transmit EIRP (per 1.25 MHz channel)
  System A (per burst)
         C-band
         L-band
	
	-0.2 dBW
	15 dBW

System B
         C-band
         L-band
	
	-5.0 dBW
	13.8 dBW

System G/T
         C-band
         L-band
         depending on beam
	
	-20.4 dB/K
	-13.2 to -11.0 dB/K

In-orbit mass	232 kg

Orbit parameters
  Altitude
  Eccentricity
  Inclination
	
	1389 km
	0 (circular)
	47 degrees

The antenna arrays are patched through a beam-forming device, which compensates for the differences in path loss over a beam gain contour. This alleviates the so-called near-far problem, described below.

There is a difference in path loss of 8 dB for a user located directly below a satellite (750 nm) and one located at an elevation of 20 degrees (1870 nm). In this case, the gain contour of the two inner spot beams (see Figure 14.11) includes a difference in gain of nearly 8 dB to compensate for the additional path loss. The middle beams compensate for extreme path loss differences, and similarly, the middle and outer beams are designed to accommodate less extreme path loss problems.

14.3.6 Gateway Segment

The gateway stations will provide the interface between the user and the rest of the world. Network Coordination gateways will control the call setup and coordination, and standard

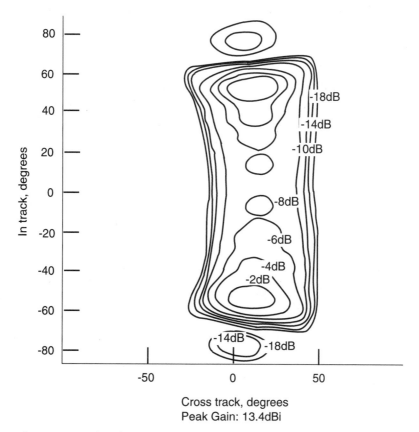

Figure 14.11 L-band Gain Contour of Middle Beams

gateways will provide the interface to the local cellular networks. Once at the cellular gateway, access to local, long-distance PSTNs or private networks is possible; refer to Figure 14.8 for system operations.

The gateway characteristics are described in Table 14.4. The ground station is a parabolic reflector with sufficient radio frequency (RF) equipment complexity to communicate with up to three satellites simultaneously (these are the satellites whose footprints cover the location of the gateway facility).

Since the Globalstar system is designed as a non-bypass system, one of the most important functions of a gateway station is the Terrestrial System Interface. The functions performed include PSTN signaling system R1 transfers, signaling system 7, in-band tones and call progress conditions for PSTN users, digitally switched PCM for gateway data channels and PSTN channels.

14.3.7 Mobile Subscriber Segment

The services Globalstar proposes are as follows:

- Radio Determination Service, position location and messaging
- Voice and data service with connection directly to and from PSTN
- Voice and data service with connections to and from private networks

The initial systems call for three types of user equipment: hand-held or vehicle RDSS units; hand-held voice units; vehicle voice units. Table 14.5 contains the characteristics of the user equipment. The voice coding bit rates are variable and, depending on selection or situation, range from 1.2 kbps to 9.6 kbps.

TABLE 14.4 Gateway Characteristics

Frequency Bands:	
Transmit	6.484–6.5415 GHz
Receive	5.1585–5.216 GHz
Ground-tracking antenna diameter	2 meters
Gain	40.2 dBi @ 6.5 GHz
Maximum sidelobe level	20 dB below main beam peak
Pointing angle range	360 degrees azimuth +5 to 90 degrees elevation
Ground acquisition antenna	Parabolic antenna
Transmitter EIRP (per burst per 1.25 MHz)	32.2 to 44.2 dBW
Receiver G/T	14.6 dB/K

TABLE 14.5 IUser Unit Specifications

Voice coding	Variable rate: 1.2 to 9.6 kbps
Bit rate	28.8 kbps
Spreading bandwidth	1.25 MHz
Error Correction Coding	Convolutional, rate 1/3, k=9
Modulation	4-phase (QPSK)
Frequency band (System A)	1610.0–1626.5 MHz
Antenna	
Type	Quadrifilar Helix or patch
Gain	1 to 3 dBi depending on elevation angle
Area of coverage	360 degrees azimuth From 5 to 90 deg elevation
Transmitter power	Peak 10 W power controlled Average 0.67 W
Receiver G/T	from -21.7 to -23.7 dB/K depending on elevation angle
Nominal Eb/No	3.5 dB

14.4 GLOBALSTAR COMMUNICATION TECHNOLOGY

Qualcomm's founding was based on the belief that there was an opportunity and ability to make major advances in the way the radio spectrum is used. To this end, direct sequence CDMA was chosen for the Globalstar system. The result is their investment in CDMA cellular technology. CDMA has some inherent advantages, including the facts that only the RF side of the equipment needs changing from existing systems and that the RF architecture is simple. A single RF transmitter is required for each CDMA rack of equipment. The complex work is done on the baseband side. The subscriber unit can be built using as few as 4 ASICs, and the cost of these units will continually decrease. Much of the complexity of the RF sections has been transferred to the modulator section.

14.4.1 CDMA Multiple Access Technique in Globalstar

How does CDMA work on satellites? The answer needs to be prefaced with a brief summary of preceding technologies as elaborated on previously. In early radio, the entire spectrum was used to communicate at very short distances. Eventually, tuned circuits were used to improve range and to allow multiple users by dividing up the spectrum in which users transmitted and received. This begot FDMA (Frequency Division Multiple Access) as a technique to separate different communication channels. With the advent of digital technology, Time Division Multiple Access (TDMA) techniques were found to be more efficient.

The spectral efficiency of TDMA and FDMA is inherently not good. FDMA requires the transmitted occupied bandwidth to be, at a minimum, equal to the required bandwidth of the communication. In reality, it required more bandwidth to reduce noise levels and provide acceptable carrier-to-noise ratios. For example a 5-kHz cellular voice channel typically requires 30 kHz of radio spectrum. TDMA has the same problems. Increasing the number of channels requires additional bandwidth. Careful parameter selection, digital reduction techniques, and reducing the frequency reuse factor from 7 to 6 or lower could also increase efficiency. The rule of thumb is that switching your cellular system to TDMA from FDMA would provide a 3-to-1 increase in the number of communication channels available.

14.4.2 Direct Seqence Spread Spectrum

In direct sequence spread spectrum (DS-SS), the signal is modulated onto a wide bandwidth spread spectrum carrier that is produced by a phase modulation of the sinusoid by a pseudo-random noise (PN) binary sequence of digital signaling elements called chips. While the resulting channel bandwidth is much larger than the data transmission rate, the benefits rely on a statistical averaging principle known as the "Law of Large Numbers."

Each signal has a different PN sequence. The desired signals are separated by a correlator in the receiver, which is matched to the desired signal's PN. The signals look like random thermal noise and are about 10–20 dB weaker than thermal noise. The correlator provides processing gain to increase the carrier-to-interference (C/I) by the ratio of the spread spectrum bandwidth to the transmission data rate.

In DS-SS, a carrier wave is modulated with a data signal x(t), and then the data modulated signal is again modulated with a high-speed wideband spreading, g(t).

If the data modulated carrier has power P, frequency w_o, and data phase $\Theta_x(t)$, then

$$s_x(t) = \sqrt{2P} \ \cos\,[w_o t + \Theta_x(t)] \qquad\qquad 14.8$$

After modulation with g(t),

$$s_x(t) = \sqrt{2P} \ \cos\,[w_o t + \Theta_x(t) + \Theta_g(t)] \qquad\qquad 14.9$$

where the phase of the transmitted carrier has two components: Θ_x due to the data and Θ_g due to the spreading signal. If BPSK modulation is used, the phase of the data changes by pi radians. So, the data modulating carrier becomes equivalent to a pulse stream of +1 or -1, and equations $s_x(t)$ can be rewritten,

$$s_x(t) = \sqrt{2P} \ x(t)\cos(w_o t) \qqua\qquad 14.10$$

where x(t) is, as before, the data signal.

If, also, the spreading signal, g(t), is BPSK, then $s_x(t)$ can be further rewritten,

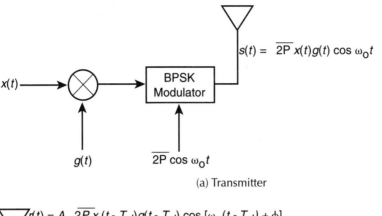

(a) Transmitter

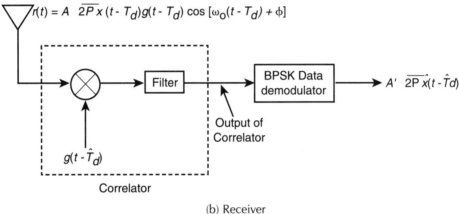

(b) Receiver

Figure 14.12 BPSK-DS Transmitter and Receiver

$$s_x(t) = \sqrt{2P}\ x(t)g(t)\cos(w_o t) \qquad\qquad 14.11$$

Figure 14.12 illustrates a simple BPSK-DS transmitter and receiver.

At the receiver, the signal is recovered by remodulating the received signal with a synchronized replica of the spreading signal $g(t - T_d)$, where signal T_d is the receiver's estimate of the transmission propagation delay. Simplistically, as shown in Figure 14.13, the correlator's output is

$$A\ \sqrt{2P}\ x(t - T_d)g(t - T_d)g(t - T_d)\cos(w_o(t - T_d) + 1) \qquad\qquad 14.12$$

where A is the system gain and 1 is a random phase angle. Since $g(t) = +/- 1$, the product $g(t - T_d)g(t - T_d)$ will be 1 if $T_d = T_d$, i.e., the receiver's code generator is synchronized exactly with the transmitter's code generator.

The pulse width of g(t) is called the code chip rate R_c. It is designed so that $R_c \gg R_b$, where R_b is the information data rate. The spread spectrum signal is basically as large as the spreading signal; thus, for a BPSK-DS spread spectrum, if the PN code chip rate is 9.6 Mchips/sec, the bandwidth is $B_t = 2 R_c$.

The spreading of the signal has made it less susceptible to interference. With the spreading, the in-band spectral density drops in proportion to R_c/R_b. So, if the R_b is 9.6 kbps and R_c is 9.6 Mchip/s, the reduction factor is $9.6 \times 10^6 / 9.6 \times 10^3 = 1000$ or 30 dB. This means that the wanted signal's power level appears to be 30 dB below its actual level and, if viewed on a spectrum analyzer, would likely appear below the noise floor. This improvement in power density is known as the processing gain G_p and is calculated as

$$G_p = R_c/R_b \qquad\qquad 14.13$$

This becomes critical to the multiple access scheme proposed, CDMA.

14.4.3 Code Division Multiple Access

In DS-CDMA, each user is given its own PN code, which is approximately orthogonal with other user codes. This will result in a cross-correlation between user codes near zero.

Figure 14.13 shows a typical DS-CDMA block diagram. Block 1 shows the data modulation of a carrier, $A\cos(\omega_0 t)$. The output of the modulator shows User 1's data waveform as $s_1(t)$, defined below,

$$s_1(t) = A_1 \cos[\omega_0 t + \varphi 1 (t)] \qquad\qquad 14.14$$

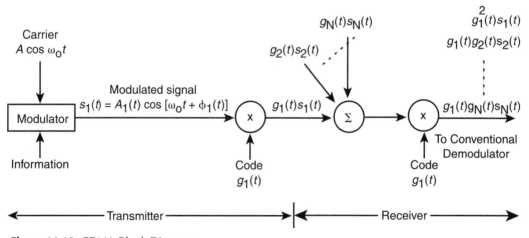

Figure 14.13 CDMA Block Diagram

This signal is multiplied by the spreading signal, $g_1(t)$, which is generated by a PN code generator. This signal $g_1(t)\, s_1(t)$ is transmitted over the channel. At the same time, all users, from 2 to N, i.e., $s_2(t)$ to $s_N(t)$, are being multiplied by their PN code spreading signal, $g_2(t)$ to $g_N(t)$. Thus, the received signal is a combination of all users.

$$g_1(t)\, s_1(t) + g_2(t)\, s_2(t) +...+ g_N(t)\, s_N(t) \qquad\qquad 14.15$$

The spreading signal $g_1(t)$ is very large compared to $s_1(t)$. Since this is so and the combined $g_1(t)\, s_1(t)$ signal's spectrum is a convolution of the $s_1(t)$ spectrum with the $g_1(t)$ spectrum, the product of $g_1(t)\, s_1(t)$ will have a bandwidth approximately equal to $g_1(t)$.

At the receiver, the signal is multiplied by $g_1(t)$, which is the PN spreading signal synchronized with User 1's received signal. The output of the receiver becomes

$$g_1^2(t)\, s_1(t) \qquad\qquad 14.16$$

plus the unwanted signals

$$g_1(t)\, g_2(t)\, s_2(t) + g_1(t)\, g_3(t)\, s_3(t) +...+ g_1(t)\, g_N(t)\, s_N(t) \qquad\qquad 14.17$$

Given the orthogonality of the codes, the desired signal becomes

$$\int_o^T g_1^2(t)dt = 1 \qquad\qquad 14.18$$

and the undesired signals result in

$$\int_o^T g_1(t)g_i(t)dt = 0 \text{ for } 1 \neq j \qquad\qquad 14.19$$

In reality, these codes are perfectly orthogonal, and the cross-correlation between codes does introduce some interference into the system which impacts the number of simultaneous users.

For a physical illustration, see Figure 14.14(a). The illustration shows the uncorrelated wideband input to the receiver. There are the wanted and unwanted signals, each spread by its own PN code, of rate R_p. The received thermal noise is also shown as a flat spectrum across the band. Figure 14.14(b), illustrates the effects of cross-correlation of a receiver with a synchronized spreading code on the desired and undesired signals. The desired signal is increased in level by the spreading processing gain. The demodulator has an input filter bandwidth wide enough to accommodate only the unspread desired signal's bandwidth.

The unwanted carriers remain spread by $g_1(t)g_i(t)$, and only that portion of the spectrum falling within the bandpass filter of the demodulator will cause interference into the desired signal.

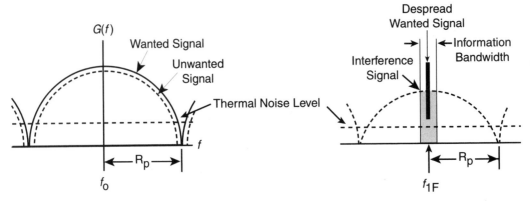

Figure 14.14(a) Spectrum at Input to Receiver, (b) Spectrum After Correlation with Correct and Synchronized PN Code.

14.5 CONCLUSIONS

While this chapter has presented an argument for the benefits of CDMA technology in the cellular, specifically, the Mobile Satellite, environment, the realities of making this system work may be a very great challenge indeed. At this time there is no CDMA terrestrial cellular network nationwide. The cost of a CDMA mobile unit is still disproportionately high compared to analog or other digital mobile units. The difficulties in designing, building, and operating the proposed Globalstar system will be great. However, the technology is sound and the inevitability of Mobile Satellite Services along with ground services exist.

Appendix to Chapter 14

A.1 GLOBALSTAR FREQUENCY AND CDMA DESCRIPTION

A.1.1 System A, Plan (TDD-FD-CDMA)

Not knowing what the FCC would rule with respect to the frequency allocation for MSS, LQSS requested different frequency plans, the final design plan being the one that concurred with the FCC's rulemaking. System A plan was proposed if both the user-to-satellite and satellite-to-user links were to fall within the L-band spectrum 1610.0–1626.5 MHz.

In order to provide sufficient user capacity at L-band on System A, a multiple-access technique, known as TDD-FD-CDMA (Time Domain Duplexing, Frequency Division, Code Division Multiple Access), is used. To further add to the complexity of this technique, beam-hopping is also employed.

Figure A.1 illustrates the frequency plan for System A. The L-band spectrum is divided into 13 subbands of 1.25 MHz each. They are spaced very close together (actual bandwidth is 1.23 MHz with 0.04 MHz of guardband employed).

Figure A.2 illustrates the frequency-reuse technique. Time-domain duplexing divides a system into 60 ms frame; there are six 10 ms time slots per frame—three time slots for transmit and three for receive. Within each time slot, the user signals will transmit or receive via only two of the six spot beams per satellite. The two beams in use (i.e., 1 and 4, see Figure A.2) will be of orthogonal polarization sense. This reduces interference from other spot beams on the satellite.

Also, the spectrum is reused 2 times for transmit and receive, 3 times by beam hopping, and 24 times, 1 per satellite, throughout the network, resulting in 144 times frequency reuse (288 times when full 48 satellite constellation is in place).

This plan will not be used because the FCC has proposed that both the L-band and S-band be allocated to users in the MSS. It is presented more as an appealing, if complex, technique for improving spectral efficiency and making this MSS cost effective.

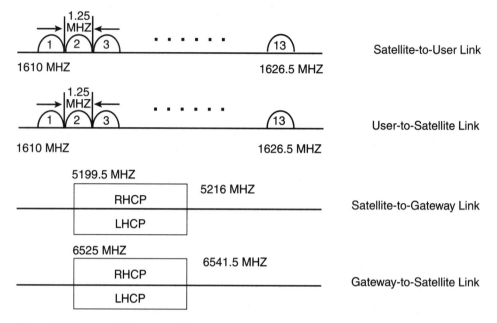

Figure A.1 System A Frequency Plan.

A.1.2 System B Plan (CDMA)

Figure A.3 illustrates the frequency plan for System B, which uses both L-band (transmit) and S-band (receive) for the user segment. As before, 1.25–MHz bandwidths are employed, but only CDMA techniques are used, as compared to TDD-FD-CDMA, since double the spectrum requires no-time-domain duplexing or beam hopping (see Figure A.4).

A.1.3 Evaluation of the Globalstar CDMA System

The efficiency of the Globalstar system is its effective frequency reuse scheme. Also, this system exploits other techniques to obtain high spectral efficiency. These other techniques consist of the following:

- Voice activity
- Error detection and correction
- Efficient modulation
- Antenna directivity (spot beams)
- Multiple satellite
- Polarization reuse
- Soft hand-off

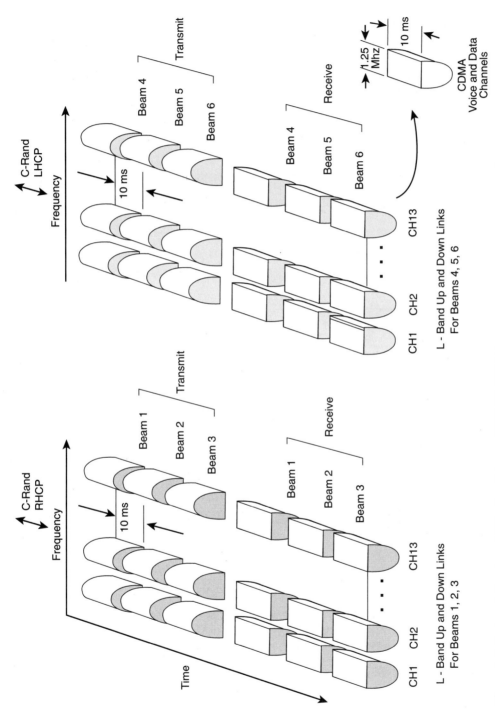

Figure A.2 System A Spectrum Utilization Scheme (TDD-FD-CDMA).

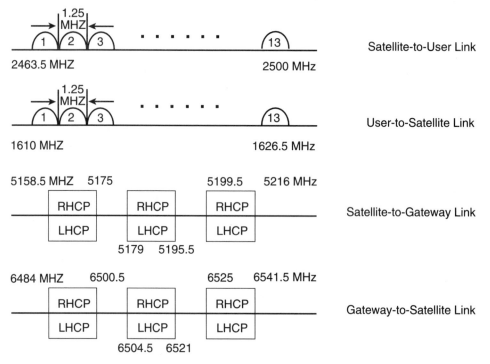

Figure A.3 System B Frequency Plan.

The major factors that determine the CDMA performance consist of the following:

- Processing gain
- Required Eb/No
- Voice duty cycle
- Frequency reuse efficiency
- Number of sectors in the cell antenna
- Number of satellites
- Polarization reuse efficiency

The details of each of these factors are discussed below.

Processing Gain: As explained above, the processing gain is the ratio of the spread bandwidth to the information rate of the system. If a vocoder rate of 4800 bps is considered and the spread bandwidth is 1.25 MHz, the resulting processing gain is 24 dB. Thus, the received carrier-to-interference (C/I) ratio can be as low as -20 dB for a received Eb/No of 4 dB.

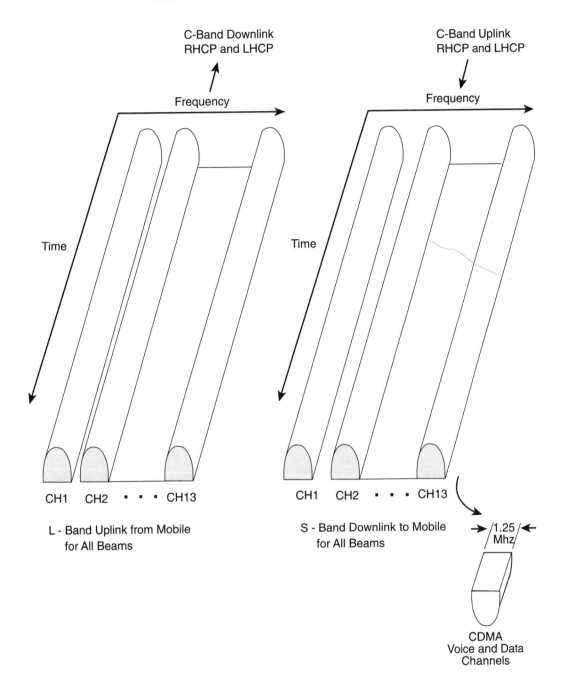

Figure A.4 System B Spectrum Utilization.

Required Eb/No: LQSS has stated that because of the efficient modulation technique and the error correction used, an Eb/No of only 3.5 is required for good performance. A power control system is used to ensure that the received Eb/No is adequate but not greater than required.

Voice Duty Cycle: Since voice signals on a typical full-duplex system are intermittent with a duty cycle of 3-to-8, capacity can be increased by an amount inversely proportional to this ratio by squelching transmission during quiet periods of each speaker.

Globalstar proposes that for a 4800 bps voice quality service, the vocoder rate is allowed to vary frame to frame, where each frame is a 20-ms period. When no activity is detected, the vocoder rate drops from 1200 bps, allowing background ambiance and synchronization signals to be preserved. As voice activity is detected, the rate jumps back to 4800 bps (or 9600 bps if that quality of service is being used). This rate switching is transparent to the user. When the lower rate is used, the transmitted power is reduced, thus reducing the system-wide interference and admitting more users.

Frequency Reuse: A property of CDMA is its ability to have a one-cell frequency reuse scheme, as opposed to a seven-cell reuse scheme for FDMA and TDMA systems. For the Globalstar system, neighboring satellite footprints will frequently overlap. This does not present a problem, and the overlap can be exploited for path diversity and increased downlink signal power.

Sectorization Gain: Globalstar satellites use an array of six spot beams. This can be compared to sectorization of cells in a terrestrial cellular system. In TDMA or FDMA systems, adjacent spot beams are not directional enough to allow frequency reuse. With CDMA, adjacent beams can reuse the same frequency, with only a small amount of capacity reduction caused by interference from adjacent beams.

Polarization Reuse: A mobile terminal antenna has a cross-polarization of only isolation 6 dB. This amounts to a C/I improvement of 6 dB, which to this CDMA system will result in an increase in system capacity of 60 percent. This is limited by the satellite's downlink power and will become realizable as all the satellites in the constellation are brought on-line and an increase in the total downlink power, through beam overlapping, occurs.

Multiple Satellites: It is anticipated by the designers of Globalstar that nearly 100 percent of the United States coverage will be achieved with 24 satellites, and nearly 100 percent coverage worldwide with 48 satellites. Most high-use areas will see double coverage 100 percent of the time. These areas, because of the CDMA technology, will have their capacity significantly increased. This increase comes about for the following reasons:

- Double coverage increases the satellites' flux density by two, increasing capacity by 70 percent.
- The satellite design does not fix the RF power for individual spot beams. If all of the traffic is concentrated in an area covered by one spot beam, all the RF power can be used by that beam.

- Double coverage allows coherent diversity combining of two received signals, thus allowing link margin to be reduced, lowering required downlink power, which, again, allows more users.

Soft Hand-off: The same channel is used in two overlapping beams. This allows "soft" hand-offs to easily occur. The call's signal is passed through a new satellite at the same time as it is passing through the original satellite. This process is a "make-before-break" process and is highly reliable because a new link is established before the old one is torn down.

A.2 LEOS AND THE FCC

In its Notice of Proposed Rulemaking (NPRM) of February 18, 1994, the FCC amended its rules pertaining to Mobile Satellite Service in the 1610–1625.6 and 2483.5–2500 MHz frequency bands. In summary, the rulemaking proposes allowing MSS operators to use both the L- and S-bands—L-band for the user-to-satellite link and S-band for the satellite-to-user. Specifically, it has determined that 11.5 MHz of the L-band spectrum will be set aside for CDMA systems and 5 MHz will be allocated for FDMA/TDMA systems (i.e., Motorola's Iridium). The impact on this is, as noted earlier, that a variation of System B will be used. This will also reduce the number of frequency subbands from 13 to 9. The time domain duplexing (TDD) technique described, while technically interesting and compelling, will not be necessary or desired because of its costs and inherent complexity.

REFERENCES

[1] Kim, K. I. "CDMA Cellular Engineering Issues," IEEE Transactions on Vehicular Technology. Aug. 1993: 42 (3): 345–349.

ADDITIONAL REFERENCES

Couch, L. W., II. *Digital and Analog Communications Systems*. (New York: Macmillan, 1993).

Jung, P., P. W. Baier, and A. Steil, "Advantages of CDMA and Spread Spectrum Techniques over FDMA and TDMA in Cellular Mobile Radio Applications," IEEE Transactions on Vehicular Technology, Vol. 42, No. 3, pp. 357–363, Aug. 1993.

Lee, W. C. Y. *Mobile Cellular Telecommunications Systems*. (New York: McGraw-Hill, 1989).

Whipple, D. P. "North American Cellular CDMA," Hewlett-Packard Journal, pp. 90-97, Dec. 1993.

Index